GIS

GIS

2nd Edition

**by Jami Dennis
with Michael N. DeMers**

GIS For Dummies®, 2nd Edition

Published by: **John Wiley & Sons, Inc.**, 111 River Street, Hoboken, NJ 07030-5774, www.wiley.com

Copyright © 2025 by John Wiley & Sons, Inc. All rights reserved, including rights for text and data mining and training of artificial technologies or similar technologies.

Media and software compilation copyright © 2025 by John Wiley & Sons, Inc. All rights reserved, including rights for text and data mining and training of artificial technologies or similar technologies.

Published simultaneously in Canada

No part of this publication may be reproduced, stored in a retrieval system or transmitted in any form or by any means, electronic, mechanical, photocopying, recording, scanning or otherwise, except as permitted under Sections 107 or 108 of the 1976 United States Copyright Act, without the prior written permission of the Publisher. Requests to the Publisher for permission should be addressed to the Permissions Department, John Wiley & Sons, Inc., 111 River Street, Hoboken, NJ 07030, (201) 748-6011, fax (201) 748-6008, or online at http://www.wiley.com/go/permissions.

The manufacturer's authorized representative according to the EU General Product Safety Regulation is Wiley-VCH GmbH, Boschstr. 12, 69469 Weinheim, Germany, e-mail: Product_Safety@wiley.com.

Trademarks: Wiley, For Dummies, the Dummies Man logo, Dummies.com, Making Everything Easier, and related trade dress are trademarks or registered trademarks of John Wiley & Sons, Inc. and may not be used without written permission. All other trademarks are the property of their respective owners. John Wiley & Sons, Inc. is not associated with any product or vendor mentioned in this book.

LIMIT OF LIABILITY/DISCLAIMER OF WARRANTY: THE PUBLISHER AND THE AUTHOR MAKE NO REPRESENTATIONS OR WARRANTIES WITH RESPECT TO THE ACCURACY OR COMPLETENESS OF THE CONTENTS OF THIS WORK AND SPECIFICALLY DISCLAIM ALL WARRANTIES, INCLUDING WITHOUT LIMITATION WARRANTIES OF FITNESS FOR A PARTICULAR PURPOSE. NO WARRANTY MAY BE CREATED OR EXTENDED BY SALES OR PROMOTIONAL MATERIALS. THE ADVICE AND STRATEGIES CONTAINED HEREIN MAY NOT BE SUITABLE FOR EVERY SITUATION. THIS WORK IS SOLD WITH THE UNDERSTANDING THAT THE PUBLISHER IS NOT ENGAGED IN RENDERING LEGAL, ACCOUNTING, OR OTHER PROFESSIONAL SERVICES. IF PROFESSIONAL ASSISTANCE IS REQUIRED, THE SERVICES OF A COMPETENT PROFESSIONAL PERSON SHOULD BE SOUGHT. NEITHER THE PUBLISHER NOR THE AUTHOR SHALL BE LIABLE FOR DAMAGES ARISING HEREFROM. THE FACT THAT AN ORGANIZATION OR WEBSITE IS REFERRED TO IN THIS WORK AS A CITATION AND/OR A POTENTIAL SOURCE OF FURTHER INFORMATION DOES NOT MEAN THAT THE AUTHOR OR THE PUBLISHER ENDORSES THE INFORMATION THE ORGANIZATION OR WEBSITE MAY PROVIDE OR RECOMMENDATIONS IT MAY MAKE. FURTHER, READERS SHOULD BE AWARE THAT INTERNET WEBSITES LISTED IN THIS WORK MAY HAVE CHANGED OR DISAPPEARED BETWEEN WHEN THIS WORK WAS WRITTEN AND WHEN IT IS READ.

For general information on our other products and services, please contact our Customer Care Department within the U.S. at 877-762-2974, outside the U.S. at 317-572-3993, or fax 317-572-4002. For technical support, please visit https://hub.wiley.com/community/support/dummies.

Wiley publishes in a variety of print and electronic formats and by print-on-demand. Some material included with standard print versions of this book may not be included in e-books or in print-on-demand. If this book refers to media that is not included in the version you purchased, you may download this material at http://booksupport.wiley.com. For more information about Wiley products, visit www.wiley.com.

Library of Congress Control Number available from the publisher.

ISBN 978-1-394-31835-3 (pbk); ISBN 978-1-394-31837-7 (ebk); ISBN 1-394-31836-0 (ebk)

Contents at a Glance

Introduction 1

Part 1: GIS: Geography on Steroids 5
CHAPTER 1: Understanding GIS 7
CHAPTER 2: Learning GIS Core Concepts 19
CHAPTER 3: Reading, Analyzing, and Interpreting Maps 37

Part 2: Geography Goes Digital 55
CHAPTER 4: Creating a Conceptual Model 57
CHAPTER 5: Understanding the GIS Data Models 69
CHAPTER 6: Keeping Track of Attribute Data 85
CHAPTER 7: Collecting Geographic Data 101

Part 3: Retrieving, Counting, and Characterizing Geography 119
CHAPTER 8: Exploring the World through Raster Data 121
CHAPTER 9: Finding Features in Vector Data 133
CHAPTER 10: Searching for Geographic Objects, Distributions, and Groups 151

Part 4: Analyzing Geographic Patterns 167
CHAPTER 11: Measuring Distance 169
CHAPTER 12: Working with Statistical Surfaces 183
CHAPTER 13: Exploring Topographic Surfaces 199
CHAPTER 14: Working with Networks 213
CHAPTER 15: Exploring Map Overlay 233
CHAPTER 16: Mastering Map Algebra and Cartographic Models 245

Part 5: GIS Output and Application 267
CHAPTER 17: Generating Output with GIS 269
CHAPTER 18: Automating GIS 293
CHAPTER 19: GIS in Organizations and Career Development 303

Part 6: The Part of Tens 317
CHAPTER 20: Ten GIS Software Options 319
CHAPTER 21: Ten GIS Data Sources 329
CHAPTER 22: Ten GIS Trends to Watch 337

Index 345

Table of Contents

INTRODUCTION 1
    About This Book. 1
    Foolish Assumptions. 2
    Icons Used in This Book 2
    Beyond the Book. 3
    Where to Go from Here 3

PART 1: GIS: GEOGRAPHY ON STEROIDS 5

CHAPTER 1: Understanding GIS 7
    Grasping the Power of GIS. 8
        Evolving from land-use tracking to spatial data science 8
        Collaborating in the cloud 8
        Transforming data into action. 9
    Meeting the GIS Collective 11
        Computer hardware 11
        Geographic software. 12
        Geographic data 13
        Methods 15
        People 16
    Thinking Spatially: Why Geography Matters 16
        Recognizing the spatial nature of analysis. 17

CHAPTER 2: Learning GIS Core Concepts 19
    Digging into the Foundations of Spatial Data 20
        Understanding spatial data types. 20
        Acquiring your spatial data 21
        Discovering how you use spatial data 22
    Understanding Maps in GIS 22
        Seeing how maps represent spatial data. 22
        Reviewing types of maps 22
        Interpreting symbols. 26
        Getting the scale right. 27
    Flattening the Earth. 29
        Measuring the Earth 29
        Understanding map projections 32
        Using coordinate systems 34
        Defining land boundaries and ownership 35

CHAPTER 3: Reading, Analyzing, and Interpreting Maps 37
    Making Sense of Symbols. 38
        Categorizing the space on a map 38
        Understanding levels of measurement 39
        Matching symbols with data types 40
    Recognizing Patterns 42
        Identifying random patterns 43
        Finding clustered patterns 43
        Observing uniform patterns 44
        Seeing patterns among diverse features 45
        Describing patterns with linear features 45
        Understanding the repeated sequence of shapes 46
    Analyzing and Quantifying Patterns 48
        Knowing your geometry and patterns 48
        Using GIS for the analysis 49
        Determining the type of pattern 50
        Choosing an appropriate tool 52
    Interpreting the Results and Making Decisions 53

PART 2: GEOGRAPHY GOES DIGITAL 55

CHAPTER 4: Creating a Conceptual Model 57
    Helping Computers Read Maps. 58
    Embracing the Model-Creation Process. 58
    Defining Your Map's Contents 59
        Choosing the data 59
        Applying the methodology 60
        Breaking down the data 61
        Verifying data characteristics 63
    Converting Maps into Digital Formats 63
        Deciding how to represent your data 63
        Weighing the benefits: Raster versus vector 66

CHAPTER 5: Understanding the GIS Data Models 69
    Examining Raster Models and Structure 70
        Representing dimension when everything is square 71
        Making a quality difference with resolution 72
        Finding objects by coordinates 72
        Finding grid cells by category 74
        Working with map layers 74
        Linking objects with attributes 75
    Exploring Vector Representation 76
        Moving beyond a simple model 76
        Adding spatial smarts to GIS with topology 77

        Using shapefiles for easy data exchange.......................79
        Building smarter systems with object-oriented models.........81
    Dealing with Surfaces...81
        Storing surface data in a raster model.......................82
        Representing surfaces in a vector model......................83
        Choosing the right model for surfaces........................84

CHAPTER 6: Keeping Track of Attribute Data........................85
    Knowing the Importance of Attribute Data in GIS...................86
        Exploring the evolution of geospatial data...................86
        Understanding the role of attribute data in GIS..............87
    Working with Tables and Database Management Systems........87
        Structuring relational data..................................88
        Connecting data with joins and relationships.................90
        Managing attributes with spatial databases...................91
        Exploring pixel-level attributes.............................92
        Searching with SQL in GIS....................................94
    Understanding Object-Oriented and NoSQL Systems...............95
        Enhancing descriptive information with object orientation......97
        Exploring emerging data storage systems.....................97
        Leveraging geodatabases and industry models for success......98
    Incorporating Data Interoperability, Standards, and Security........99
        Integrating AI and machine learning in GIS...................99
        Ensuring interoperability with standards.....................99
        Addressing security and privacy in GIS......................100

CHAPTER 7: Collecting Geographic Data............................101
    Identifying Quality Data..101
        Understanding why quality matters...........................102
        Evaluating factors for collecting high-quality data..........102
    Importing Data...103
        Collecting data with a GPS receiver.........................103
        Using remote sensing to collect data........................106
        Comparing passive versus active remote sensing..............106
        Enhancing and classifying images............................107
        Working with samples of data................................108
    Geocoding Data..109
        Turning addresses into points...............................110
        Creating points and lines by reference......................111
    Accessing Data with API Connections.............................112
        Understanding APIs..112
        Fetching data from APIs through scripting...................113
    Building Data about the Data......................................114

PART 3: RETRIEVING, COUNTING, AND CHARACTERIZING GEOGRAPHY ... 119

CHAPTER 8: Exploring the World through Raster Data ... 121
Identifying and Locating Features in Raster Data ... 121
    Locating areas of interest ... 122
    Analyzing linear feature ... 124
    Exploring areas and distributions ... 125
Performing Searches in Raster Data ... 126
    Searching in simple rasters ... 127
    Searching DBMS and cloud-supported rasters ... 127
    Using machine learning and AI for searches ... 128
Calculating Statistics and Summarizing Data ... 128
    Getting simple statistics ... 129
    Exploring advanced statistical techniques ... 130
    Visualizing and interpreting raster data ... 130
    Reporting and sharing results ... 131

CHAPTER 9: Finding Features in Vector Data ... 133
Getting Explicit with Vector Data ... 134
Seeing How Data Structure Affects Retrieval ... 135
    Building relationships ... 135
    Optimizing your data for speedy searches ... 135
Choosing the Right Data Source ... 136
    Targeting the right data ... 136
    Predicting the outcome ... 138
Locating Specific Features with SQL ... 139
    Using SQL queries based on attributes ... 139
    Exploring advanced SQL functions ... 142
    Making the most of your search results ... 142
Searching Vector Data with Geography ... 144
    Using spatial operators ... 145
    Finding features based on distance or buffers ... 146
Counting, Tabulation, and Summary Statistics ... 147
    Performing basic statistical analysis ... 148
    Visualizing and summarizing data ... 148
Validating and Verifying the Results ... 149

CHAPTER 10: Searching for Geographic Objects, Distributions, and Groups ... 151
Searching Polygons in a GIS ... 152
Searching for the Right Objects ... 153
    Extracting specific information with attribute searches ... 153
    Understanding polygon metrics ... 154

      Analyzing point distributions............................157
         Grouping and ranking data..........................158
   Locating Map Features......................................159
      Searching by attributes..............................160
      Searching by shape and size.........................160
      Searching by proximity..............................162
      Searching by groups and clusters....................162
      Combining multiple search methods...................163
   Defining the Groups You Want to Find........................164
      Grouping by common properties.....................164
      Grouping by location and patterns..................165
      Grouping by what you already know.................166

PART 4: ANALYZING GEOGRAPHIC PATTERNS 167

CHAPTER 11: Measuring Distance 169
   Taking Absolute Measurement...............................170
      Finding the shortest straight-line path...............170
      Measuring Manhattan distance.......................172
      Calculating distance along networks................174
   Establishing Relative Measurements 175
      Adjacency and nearness.............................175
      Separation and isolation 177
      Containment and surroundedness....................177
   Measuring Functional Distance..............................178
      Navigating non-uniform surfaces (anisotropy).......179
      Considering the intangibles........................180
      Creating the functional surface....................181
      Calculating the functional distance................181

CHAPTER 12: Working with Statistical Surfaces 183
   Examining the Character of Statistical Surfaces..............183
      Understanding discrete and continuous surfaces.....185
      Exploring rugged and smooth surfaces...............186
      Climbing steep surfaces.............................187
      Determining slope and orientation..................187
   Working with Surface Data.................................188
      Collecting and preparing surface data..............189
      Sampling statistical surfaces......................189
      Displaying and analyzing Z values..................190
      Challenging the rules of continuous data 191
   Predicting Values with Interpolation 191
      Determining values with linear interpolation.......192
      Using non-linear interpolation 194
      Estimating with distance-weighted interpolation....195
      Exploring beyond basic techniques..................196

CHAPTER 13: Exploring Topographic Surfaces ... 199

- Modeling Visibility with Viewsheds ... 200
  - Using viewsheds in the real world ... 200
  - Simulating line-of-sight with ray tracing ... 201
- Mapping Watersheds and Basins ... 203
  - Understanding watersheds ... 203
  - Working with basins in GIS ... 204
- Characterizing Water Flow ... 205
  - Understanding why flow direction matters ... 205
  - Modeling flow direction in GIS ... 205
  - Determining flow speed in GIS ... 207
- Defining Streams ... 208
  - Understanding stream networks in watersheds ... 208
  - Locating streams in GIS ... 209
  - Identifying methods that work for you ... 210

CHAPTER 14: Working with Networks ... 213

- Measuring Connectivity ... 213
  - Recognizing why connectivity matters ... 214
  - Calculating the gamma index ... 214
- Working with Impedance Values ... 217
  - Understanding why traffic is good or bad ... 218
  - Modeling impedance in traffic networks ... 218
- Navigating One-Way Paths ... 221
  - Understanding one-way systems ... 221
  - Incorporating one-way paths in network models ... 222
- Defining Circuitry ... 223
  - Understanding how circuits improve networks ... 223
  - Measuring and modeling circuits ... 224
- Working with Turns and Intersections ... 224
  - Recognizing the importance of turns and intersections ... 224
  - Encoding and using turns and intersections ... 225
- Directing Traffic and Exploiting Networks ... 226
  - Finding the shortest path ... 227
  - Finding the fastest path ... 227
  - Finding the nicest path ... 228
  - Defining service areas ... 229
- Working with Networks in GIS ... 231

CHAPTER 15: Exploring Map Overlay ... 233

- Exploring Basic Overlay Methods ... 234
  - Selecting by location ... 234
  - Performing spatial joins ... 235

Using Advanced Overlay Techniques 236
    Performing union overlays.............................. 236
    Using intersection overlays 237
    Applying identity overlays 238
    Applying clip overlays 239
    Exploring symmetrical difference overlays 241
Understanding Raster Overlay 242

CHAPTER 16: Mastering Map Algebra and Cartographic Models ... 245

Creating Cartographic Models................................. 246
Understanding Map Algebra 247
Performing Map Algebra Functions 248
    Exercising control in cartographic modeling............. 249
    Using local functions................................... 249
    Using neighborhood functions 251
    Exploring zonal functions............................... 254
    Resampling raster data.................................. 257
    Using global functions.................................. 258
Building a Model 260
    Planning your analysis 261
    Preparing your data inputs 262
    Implementing your model 262
Testing a Model 264
    Verifying that the model works.......................... 264
    Assessing the results................................... 265
    Gauging whether your model makes sense 265
    Meeting end-user needs.................................. 266

PART 5: GIS OUTPUT AND APPLICATION ... 267

CHAPTER 17: Generating Output with GIS ... 269

Exploring Cartographic Design 270
    Reviewing the design process............................ 270
    Understanding color theory.............................. 271
    Choosing appropriate fonts.............................. 272
    Designing for accessibility 273
Mapping Data.. 275
    Mapping qualitative data 275
    Mapping quantitative data............................... 276
    Creating classes.. 276
    Stepping up your classification game.................... 278
Laying Out Your Map 279
    Using essential map elements............................ 279
    Factoring in graphic design 280

Optimizing maps for different media..........................282
Developing Interactive Maps and Apps........................283
    Choosing tools and platforms for web mapping.............283
    Enhancing maps with interactive features.................284
    Creating data dashboards.................................284
    Developing GIS-based web applications....................285
    Incorporating user interaction and feedback..............286
Creating Noncartographic Outputs.............................287
    Automating reports and data exports......................287
    Producing lists and summary statistics...................288
    Creating systems for real-time monitoring and alerts.....289
Testing and User Feedback....................................289
    Conducting usability testing.............................290
    Gathering feedback from diverse audiences................290
    Iterating based on user input............................291

CHAPTER 18: Automating GIS.................................293
Getting to Know GIS Programming Languages....................293
Making GIS Work Smarter with Scripts.........................296
Getting Data through APIs and Web Services...................297
Understanding Best Practices.................................298
    Reusing the code...299
    Handling errors..299
    Optimizing performance...................................300
    Documenting the process..................................300

CHAPTER 19: GIS in Organizations and Career Development....303
Transforming Organizations with GIS..........................304
    Exploring GIS roles and interactions.....................304
    Adapting GIS to organizational needs.....................305
Designing and Implementing GIS in Your Organization..........307
    Aligning GIS with organizational goals...................307
    Overcoming integration challenges........................308
    Managing people problems.................................308
    Planning for integration.................................309
Building a Career in GIS.....................................311
    Understanding what it takes to succeed in GIS............311
    Learning and growing in your GIS career..................312
    Networking and professional growth.......................313
    Exploring career paths in GIS............................315

PART 6: THE PART OF TENS......317

CHAPTER 20: Ten GIS Software Options......319
ArcGIS Pro......320
QGIS......320
MapInfo Pro......321
GRASS GIS......322
AutoCAD Map 3D Toolset......323
Google Earth Engine......324
GeoServer......325
ERDAS IMAGINE......325
Global Mapper......326
OpenCities Map......327

CHAPTER 21: Ten GIS Data Sources......329
OpenStreetMap......330
ArcGIS Online and Living Atlas......331
Natural Earth......331
IPUMS NHGIS......332
Copernicus Data Space Ecosystem......332
USGS EarthExplorer......333
SEDAC......333
WorldClim......334
GeoPlatform.gov......334
INEGI......334

CHAPTER 22: Ten GIS Trends to Watch......337
Integration of AI and Machine Learning......337
3D GIS and Augmented/Virtual Reality......338
Cloud-Based GIS and SaaS Solutions......339
Expansion of Remote Sensing Capabilities......340
Evolving Geospatial Data Privacy and Ethics......341
Growth of Open-Source GIS and Data......341
Spatial Digital Twins......342
Advances in Location-Based Services......343
Citizen Science and Crowdsourcing......343
Geospatial Education and Workforce Development......344

INDEX......345

Introduction

Are you trying to figure out GIS because your instructor assigned it, your boss expects you to use it, or your organization already has it in place? Maybe you've heard GIS mentioned in meetings and need to understand what exactly it does. Or maybe you just want to make better maps and analyze geographic data more effectively.

No matter how you landed here, I designed this book to get you comfortable with GIS without overwhelming you. Whether you're using GIS for the first time, need a refresher, or want to understand what it can do for you, you're in the right place!

About This Book

Unlike many books on GIS, this one isn't meant to keep you spellbound for days or weeks. Instead, you can use this book when you need to answer basic questions or figure out what questions to ask your GIS friends. Think of this book as a reference you can use to find what you need when you need it.

The book gives you a big picture look at GIS — everything from the parts that make up GIS (see Chapter 1) to gathering data (see Chapter 7) to careers in GIS (see Chapter 19), and much more. Wherever your interests in GIS point you, find those topics in the Table of Contents or Index and jump right in.

GIS terminology can sometimes be confusing, but don't worry; I explain terms as they come up. That said, I want to point out a few conventions used throughout this book. The term *raster* refers to a GIS data structure composed of square grid cells. The term *vector* refers to the data structure made up of points, lines, and polygons. The word *spatial* simply means anything related to location or the arrangement of things in space.

A term, I'm about to define appears in *italics*. Also, I show URLs and code terms in `monofont` typeface to set them apart from the regular text.

Foolish Assumptions

This book assumes that you've heard about GIS but don't know all that much about its inner workings and hidden mechanisms. Many people think that GIS (Geographic Information System) means GPS (Global Positioning System) because more people have heard the term *GPS*. In reality, GPS is just a part of GIS, as I explain in Chapter 7.

I also assume you have something more than a casual interest in GIS, so the book covers what GIS is, what it does, and how it can help you with what you do in your organization. Here are a few other assumptions this book makes:

» **You know what a map is.** GIS relies heavily on maps and map-related data. I assume that you have used a map of some kind but aren't an expert in either making or using maps. I provide all the background you need to help you become familiar with how maps represent real-world geography.

» **You know what geography is.** I assume that you've taken a geography class at some point in your life. I don't expect you to be or think like a geographer, however, so I guide you on that path as well. After you figure out how to think like a geographer (in mapping terms), GIS can become your friend and ally. You may even find it fun to use, as I do.

» **You use some form of computer from time to time.** GIS relies on computers. I don't expect that you're a computer wiz, but I do assume that you know what data files and software programs are and how to use a computer interface. Beyond that, I explain some of the inner workings of the GIS software and databases so that you can ask GIS experts intelligent questions.

Icons Used in This Book

GIS For Dummies, 2nd Edition, uses icons that help direct your reading. These little graphics can save you time by letting you find all the high points quickly.

TIP

The Tip icon provides a few helpful hints about shortcuts, best practices, and ideas of what to try doing with GIS. These tips help you do the right things at the right time for the right reasons.

WARNING

I use the Warning icon to keep you from making mistakes that are very hard to recover from. I don't use these often, but when I do, please heed the warning!

The Remember icon highlights important information for using GIS, so take note of it and, well, remember it.

 The Technical Stuff icon indicates an extra or more advanced detail that I've included. You can skip over these tidbits if you just want to stick to the basics.

Beyond the Book

In addition to this book, check out the bonus Cheat Sheet available on the Dummies website. It gives you a quick rundown of GIS and what you can do with it, a handy guide to raster-based functions, things to know about maps (like scales, projections, and datums), and the X, Y, and Z of GIS. To get access to the Cheat Sheet, visit Dummies.com and enter **GIS For Dummies Cheat Sheet** in the search box.

Where to Go from Here

This book isn't meant to be read cover to cover (although you certainly can if you want the full experience). Each chapter stands on its own to guide you through a specific topic related to GIS. That being said, here are a few recommendations for where you might want to jump in:

» **If you're a new GIS user** or someone who had to learn GIS software for a specific task but never really grasped the key concepts, start with Chapter 2, especially the section "Flattening the Earth." Many seasoned GIS professionals struggle with map projections and coordinate systems, so getting a jump start on this essential knowledge is valuable.

» **If you're interested in how GIS data works,** check out Chapter 5 for an overview of raster and vector data models. These are the two main GIS data types, and understanding their differences will help you choose the right data type for your analysis, mapping, and spatial queries.

» **If you need to create or collect GIS data,** go to Chapter 7, which walks you through the different ways to import, collect, and geocode geographic data. This chapter is a great place to start if you need to get data into your GIS before you can analyze or map it.

>> **If you're focused on analysis,** jump into Part 4, which covers spatial analysis methods, from measuring distances (Chapter 11) to working with surfaces (Chapter 12) and networks (Chapter 14).

>> **If you're here to make maps and share your results,** head over to Chapter 17, which covers cartographic design, map layouts, and interactive mapping tools.

No matter where you start, you can use the Table of Contents to find the topics that matter most to you.

1 GIS: Geography on Steroids

IN THIS PART . . .

Discover all the parts of Geographic Information Systems (GIS) and how they work together.

Get familiar with the core concepts of GIS, including the three basic types of maps.

Find out how to identify map symbols and patterns and transform them into meaningful data analysis.

> **IN THIS CHAPTER**
>
> » Breaking down the basics of GIS
>
> » Understanding how the pieces of GIS combine to form a complete system
>
> » Practicing how to think spatially like a geographer
>
> » Exploring real-world applications of GIS

Chapter **1**

Understanding GIS

Everything you experience from day to day happens somewhere in geographic space. As a result, you can represent your world and your experiences in it by using maps. You use those maps to navigate to a restaurant, track a package, decide where to locate a store, guide conservation efforts, and satisfy hundreds of other applications where location matters. Whenever location plays a role in decision-making, GIS helps you understand patterns, relationships, and trends that shape the world around you.

GIS is a powerful tool for analyzing, visualizing, and understanding spatial relationships. It helps businesses find ideal store locations, urban planners design smarter cities, scientists track climate change, and emergency responders map disaster-relief efforts.

In today's digital world, GIS has evolved far beyond paper maps. It's interactive, dynamic, and integrated with real-time data, enabling you to analyze everything from traffic patterns to weather conditions. If you're an aspiring GIS professional, a student, or just curious about how maps shape decisions, this book will help you understand how GIS works and why it's an essential tool in so many fields.

This chapter offers you a view of the GIS landscape (ahem!) to give you a sense of what GIS encompasses, how its components work together, and how thinking spatially can help you uncover useful patterns and meaningful insights.

Grasping the Power of GIS

GIS has come a long way since its inception in the 1960s as an innovative tool for land-use management. Today's GIS bridges geography and technology, giving resource managers and decision makers the ability to harness geographic data to solve real-world problems.

Beyond creating maps, GIS is also a tool for analyzing and visualizing data. It helps you find patterns, relationships, and trends that you'd never spot in a spreadsheet. What started as a tool for land-use planning and resource management has grown into a powerhouse of new capabilities and advanced techniques, ranging from 3D visualization to real-time data analysis and artificial intelligence (AI). The evolution of GIS continues to be driven by faster computers, bigger datasets, and new ways to apply GIS across all industries.

Evolving from land-use tracking to spatial data science

GIS didn't begin with simple mapmaking but was created to solve practical problems. GIS was developed in the 1960s by Roger Tomlinson for Canada's national land-use inventory. Through his work, Tomlinson pioneered the idea of digitally storing, layering, and analyzing geographic data.

Since then, GIS has grown into a full-fledged, spatial data science tool. No longer just for tracking land use, GIS enables you to visualize data, analyze trends, and forecast future scenarios. Forestry specialists use GIS to forecast wildfire risks; climate scientists use it to simulate sea level rise; and urban planners use it to analyze urban growth patterns. Government agencies, businesses, and nonprofit organizations rely on GIS to take on some of today's biggest challenges, like climate change and disaster response, helping to monitor, predict changes, and deploy resources. GIS even plays a role in some everyday challenges, like helping me figure out where I left my smartphone.

Collaborating in the cloud

Gone are the days when GIS was stuck on a single desktop computer. Cloud-based GIS has revolutionized storing, analyzing, and sharing geographic data, making it more accessible and collaborative than ever. Instead of relying on one powerful, stand-alone machine, you can tap into the cloud (the vast network of remote computers that store and process data on the internet) to process and share data from anywhere to anywhere.

One of the most widely used GIS cloud platforms is ArcGIS Online, from Esri, which enables you to create, analyze, and share maps entirely in the cloud. Many organizations rely on ArcGIS Online for its integrated suite of tools, spatial data, and ready-to-use apps, making it a go-to solution for professional GIS work.

But ArcGIS Online isn't the only game in town. Open-source web mapping tools like Mapbox, Leaflet, and OpenLayers provide flexible, cost-effective ways to bring GIS to the web. If you're working on a shoestring budget, these options help you bring interactive maps to the web. Although these tools don't include built-in spatial analysis like ArcGIS Online, you can pair them with open-source GIS backends like PostGIS (a spatial database) and GeoServer (a web-based GIS platform) to handle more advanced geospatial tasks.

Over the last decade, open-source GIS solutions have evolved rapidly, offering free and flexible alternatives to proprietary solutions. Tools like QGIS (a powerful desktop GIS) and GRASS GIS (for advanced geospatial analysis and modeling) allow individuals and organizations to perform GIS work without high costs. Add data from OpenStreetMap (OSM), which provides freely available geographic data, and you have pretty much all you need for taking on just about any GIS project.

Technology is not the only force shaping GIS. Engaged user communities are another key driving force for GIS. Open-source GIS benefits from its collaborative development environment, in which users contribute improvements and share knowledge freely. Meanwhile, even proprietary GIS platforms like ArcGIS Online evolve based on user feedback, with strong professional networks helping to drive innovation.

The shift toward cloud-based and open-source GIS puts powerful mapping and analysis tools into more hands than ever before. Whether you're collaborating on a city planning project, tracking migratory birds across the Pacific Flyway, or building interactive web maps, these platforms make it easy to work with geographic data anywhere. So what are you waiting for? A whole world of flexible, affordable, and even free GIS tools are out there, ready for you to explore!

Transforming data into action

GIS helps people make sense of the world by bringing together geography, computing, and problem-solving. With GIS, you can input, store, retrieve, edit, analyze, and visualize geographic data. But you may wonder what that looks like in the real world, which has real-world problems to solve. Here's just a taste of the many ways in which GIS can provide solutions:

> » **Data management:** Store, retrieve, and edit geographic data, from maps to live sensor feeds (like real-time traffic data)

- **Analysis:** Count, group, classify, isolate, and measure features and their patterns across the landscape
- **Mapping:** Overlay different datasets to uncover relationships, compare features, and make new maps
- **Visualization:** Create and manipulate 2D and 3D visualizations, predict missing values, and model changes over time
- **Route optimization:** Find the shortest, fastest, or most scenic path, identify potential customers, and locate businesses
- **Topographic analysis:** Perform tasks like modeling surface flow or measuring visibility from a specific location

This list of capabilities shows the diversity and power of GIS. But its real power doesn't derive from what GIS can do but rather from how people use it. Here are some areas in which you might see GIS in action:

- **Disaster management:** Emergency responders use GIS to map wildfire perimeters, identify evacuation routes, and prioritize relief efforts based on population density and risk factors.
- **Urban development:** Planners analyze land-use patterns and infrastructure needs, combining geographic and demographic data to design smarter cities.
- **Environmental conservation:** Conservationists track deforestation, predict wildlife migration patterns, and monitor water quality using GIS-based analyses.
- **Business applications:** Retailers optimize their supply chains, identify target markets, and pinpoint ideal store locations using GIS tools.
- **Public health:** Epidemiologists map disease outbreaks, analyze health-care access, and predict future areas of concern.

This section describes just a small slice of the capability of GIS in tackling everything from disaster response to business strategy. GIS brings together time-tested geographic methods with cutting-edge computing, allowing you to analyze places, patterns, and problems in ways that were impossible just a few decades ago.

GIS is truly transformative software, reshaping the way decisions are made around the world every day, from helping first responders save lives to giving conservationists the tools they need to protect natural habitats. Beyond the software. GIS is part of a much bigger system that connects data, technology, and people. That's what this book is all about.

Meeting the GIS Collective

If someone asks, "What is GIS?" the easy answer is that it's software for making maps and analyzing spatial data. In reality, though, it's a powerful system composed of multiple parts. I like to think of it as a team (see Figure 1-1), with each member playing a necessary role. Together, what they create sometimes even seems magical. Meet the five players of Team GIS:

- **Computer hardware:** The machines and devices that process and display GIS data
- **Geographic software:** The programs that help you analyze and visualize data
- **Data:** The fuel that drives GIS (Without it, you have nothing to analyze!)
- **Methods:** The techniques and workflows that turn raw data into meaningful insights
- **People:** Professionals across countless fields to solve real-world problems

Read on to take a look at what each of these components brings to the GIS collective.

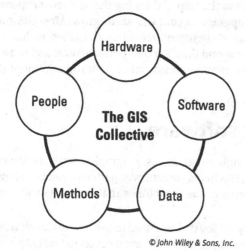

FIGURE 1-1: GIS is a collection of hardware, software, data, methods, and people.

© John Wiley & Sons, Inc.

Computer hardware

GIS software does the heavy lifting, but hardware is what makes it all possible. Modern GIS doesn't just run on desktops anymore; you find GIS on laptops,

CHAPTER 1 **Understanding GIS** 11

tablets, mobile devices, and even cloud servers. Here's a quick look at the hardware that keeps your GIS running:

» **Data collection devices:** These devices include GPS units, drones, satellite sensors, and even smartphones that gather geographic data. The data may come from a field scientist recording wildlife locations or a drone mapping a construction site, but GIS always starts with data collection. (See Chapter 7 for details on collecting geographic data.)

» **Input devices:** You can enter data directly into GIS using onscreen digitizing, scanners, and even manual digitizers (if you can find one outside a museum!).

» **Data storage and processing:** GIS deals with huge amounts of data, so having the right hard drives, cloud storage, memory, and high-performance computing makes a big difference to effective use and management of all that data.

» **Output devices:** Monitors, printers, and plotters (large-format printers) help you visualize and share your GIS work. You can display it through an interactive map on your screen or a printed map for a presentation, but GIS results need to be seen!

TIP

Before you invest in GIS software, make sure that it's compatible with your hardware. Also, don't fall into the trap of thinking that the most expensive setup is the best; you can always upgrade as your GIS needs grow. Most GIS software includes documented minimum system requirements, and often includes recommended specifications that go beyond that. For best performance, and to avoid future frustrations, start with the recommended rather than the minimum (Future You will thank you).

Geographic software

GIS software is the engine that drives geographic analysis. Beyond just storing and displaying maps, GIS helps you uncover patterns, relationships, and trends in your data. Here are some of the capabilities that make GIS software powerful:

» **Interactive analysis:** Say that you need to see how urban growth affects green spaces. GIS lets you layer different datasets to find hidden trends. (Chapters 5 and 15 explain working with layering.)

» **3D Visualization:** Modern GIS offers more than just flat maps. Many tools can now display elevation models, create fly-through animations, and even create *digital twins,* or virtual replicas of cities, buildings, or entire environments that update with real-world data.

» **Real-time data integration:** Live traffic feeds, weather updates, and satellite imagery can all be integrated into GIS, giving you up-to-the-minute insights.

Following are just some of the ways people use GIS software:

» **An environmental scientist** overlays vegetation data on a 3D terrain model to study how elevation affects plant growth.

» **A real estate agent** pulls up a map showing houses within the client's price range, complete with property details and virtual tours.

» **A city planner** presents an animated time-lapse video showing how urban sprawl has changed over the past 30 years.

» **A national park visitor** uses an interactive GIS app, taking a virtual fly-through tour of a hiking trail before visiting.

GIS software offers myriad possibilities to help you understand the world around you in new ways.

Geographic data

Hardware and software are the driving forces behind GIS analysis and products, but data provide the fuel. No data, no GIS. GIS works with many types of data, but they're easier to understand when grouped into two basic categories:

» **Primary data:** Data you collect yourself. This is data gathered firsthand for a specific purpose. Because you're in control, this data is often the most reliable and tailored to your needs. Examples include:

- **GPS field surveys:** Mapping roads, trails, or invasive plant species
- **Drone imagery:** Capturing high-resolution aerial views of construction sites, crops, or coastlines
- **Sensor readings:** Recording environmental conditions, like temperature, air quality, or water levels
- **Crowd-sourced data:** Using data contributions from citizen science projects like iNaturalist or tapping into real-time updates like OpenStreetMap contributions

» **Secondary data:** Data collected by others for unrelated tasks. Although these data are collected by someone else, they're still incredibly valuable. Examples include:

- **Census data:** Government-collected population, housing, and economic statistics
- **Satellite imagery:** Preexisting earth observation data from agencies like NASA or commercial providers
- **Weather and climate data:** Temperature trends, precipitation patterns, and severe weather events
- **Business and economic data:** Real estate prices, market trends, or customer demographics
- **Historical maps:** Old survey maps or digitized archives for land-use changes

GIS data can come from a variety of sources, including government agencies, open-source projects, private companies, or even your own organization's records. If you're not sure where to start, check out Chapter 21 for ten data sources that can get you started.

Most GIS data you collect, whether from a primary or secondary data source, will need some sort of preparation (or "cleaning") before you can use it. Here are the basic steps for getting data into your GIS:

1. **Decide what you need.**

   Your project goals will help you figure out the best data to use. For instance, determine whether you're working with existing datasets, collecting new field data, or using remote sensing (satellite or aerial imagery).

2. **Gather the data.**

   Gathering data may involve surveying using GPS or mobile GIS apps, downloading government datasets, flying a drone (uncrewed aircraft system, or UAS) to capture high-resolution imagery, or accessing satellite imagery from providers like Landsat or Sentinel.

3. **Format the data for GIS.**

   Data often needs to be converted, formatted, or cleaned up before you can use it in GIS. You need to perform some QA/QC checks on the data, clean up any errors or extraneous data items, and then save or export the data file to a compatible format. Common GIS data formats include:

   - **Shapefiles and GeoJSON:** Used for vector data (points, lines, and polygons)

- **Raster images (TIFF, JPEG2000):** Used for satellite or elevation data
- **Tabular data (CSV, Excel, database formats):** Used for attribute information

I can't include a section about data in this chapter without mentioning metadata. *Metadata* is data about data. It tells you where the data came from, how and when it was collected, how accurate it is, and other important details that let you and others know that it came from an authoritative source. If you don't know where your data came from or how old it is, be cautious, especially if you're using it for decision-making. See Chapter 7 for more about metadata.

REMEMBER

Good data equals good GIS analysis. If your data is outdated, messy, or inaccurate, your GIS results will be, too (garbage in, garbage out). So be sure to always review the metadata for the data you gather from other sources. Likewise, be sure to always include metadata with any data you create.

Methods

Methods comprise the "how" of GIS. With GIS, it's not just about *what* you analyze, it's about *how* you analyze it. The methods you use can reveal patterns, relationships, and trends that might not be obvious when looking at the raw data.

Here are some of the most common methods and how GIS users employ them:

» **Overlay analysis:** Used to stack multiple datasets (like land use, soil type, and flood zones) to see how they interact. For example, if you want to find the best place to build a school, overlay population density, zoning laws, and flood risk maps can narrow down your options.

» **Network analysis:** Good for finding the shortest route, optimizing delivery schedules, or modeling transportation systems. For instance, when you need to map the fastest emergency response routes for ambulances, network analysis enables you to factor in traffic conditions and road closures to give you the best paths.

» **Spatial interpolation:** Helpful for estimating unknown values based on surrounding data points. For example, by analyzing nearby sensor data, meteorologists use spatial interpolation to estimate temperature and rainfall in areas without weather stations.

» **Geostatistics and machine learning:** These are advanced methods that help predict patterns and trends. For example, you can use these techniques to model wildlife migration patterns, analyze crime hotspots, and predict fire risk.

CHAPTER 1 **Understanding GIS** 15

REMEMBER

Understanding how these methods work will help you figure out the right questions to ask when you're analyzing data. The right questions lead to meaningful results. You can discover much more about GIS methods in the chapters in Part 4.

People

GIS is made and used by people to help people. Across industries like business, government, military, education, nonprofits, and healthcare, GIS helps people do their work effectively. In doing so, it also enhances the way organizations work and decisions are made.

Behind every great GIS-powered decision is a system that needs to be built, maintained, and integrated into daily workflows. To be successful, organizations can't just install software and then sit back and hope for the best; they need the right people to implement it. GIS professionals serve this purpose, and depending on the organization, GIS implementation may involve these roles:

» **GIS administrators or IT specialists** who set up enterprise GIS systems

» **GIS analysts and consultants** who determine how GIS fits into the business processes and building an implementation plan around those processes

» **Geospatial architects** who design scalable solutions, often using a combination of networked workstations and cloud-based solutions

TIP

If your organization is new to GIS, working with an experienced GIS consultant or enterprise GIS specialist can help ensure a smooth transition and long-term success. Be sure to always plan and budget for training. Well-trained staff keep the whole system running smoothly, from implementation to ongoing operations and future upgrades.

Thinking Spatially: Why Geography Matters

Every tool is designed to solve a problem. Sharp things cut, heavy things hammer, and pointy things hold stuff together. GIS is a tool with a problem-solving purpose as well: to solve geographic problems.

Geographers needed a way to analyze and solve location-based problems, so they created GIS as a problem-solving toolkit to address geographic questions. Today, many different fields use GIS, but the questions that it helps answer are still fundamentally geographic. So, to get the most out of your GIS, think like a geographer.

Thinking like a geographer doesn't mean memorizing capital cities or knowing Wisconsin's best-selling cheese. Those tidbits of knowledge may come in handy on Trivia Night, but they can't help you think like a geographer. On the contrary, thinking like a geographer means seeing the world through a spatial lens, recognizing patterns, relationships, and connections based on location. When you start to see (or imagine) maps in everything, you're thinking like a geographer.

Geography affects you every single day. It encompasses the logistics that bring you cereal for breakfast, the sensors that tell you when to grab an umbrella, and the real-time traffic updates that help you steer clear of traffic jams. Think about how often you make decisions based on location. Geography plays a role in answering questions like these:

- What's the fastest way to get downtown during rush hour?
- Where's the best place to open a new clothing store?
- What restaurants are within walking distance of my hotel?
- Which neighborhood has the best schools?
- Why is cancer mortality higher in some neighborhoods than others?
- How is climate change contributing to the distribution of this bird species?
- Where will traffic congestion be worse in ten years based on population trends?

Geographers and GIS professionals ask these kinds of spatial questions every day. GIS helps them figure out the answers by enabling them to identify, characterize, question, analyze, visualize, explain, and finally apply their knowledge of patterns, distributions, and relationships.

You don't need to be a geographer to think like one. But you do need to think spatially to take full advantage of GIS.

Recognizing the spatial nature of analysis

Geographers recognize that the world is interconnected, but a core principle of geography is that places and features that are closer together tend to be more related to one another than those that are farther apart. This idea helps explain why location matters in business, transportation, public health, and more.

When analyzing problems with GIS, several spatial factors come into play. Here are a few key factors to help you start thinking spatially:

- **Density:** If you're an urban planner, the more houses an area has (the greater density), the more potential riders a public transit system has.

CHAPTER 1 **Understanding GIS**   17

- **Sinuosity:** Maybe you've noticed how winding streets force you to drive slowly. Urban planners design curvy subdivisions to reduce speeding and create safer pedestrian areas.

- **Connectivity:** Remote towns with poor road access often struggle with economic growth because goods, services, and people have a harder time moving in and out.

- **Pattern change:** As farmland and open space turn into housing developments and warehouses, local food production and wildlife habitat may pay a price.

- **Movement:** Hurricanes, migration patterns, and traffic all depend on movement over time. For example, meteorologists track hurricane paths to predict where the storms will go next, potentially saving lives.

- **Shape:** A developer looking to build a house may prefer a square-shaped lot over an awkwardly shaped one for easier construction and design.

- **Size:** Large farms need bigger parcels of land for efficient production. Plus, large farm equipment doesn't work well on tiny fields.

- **Isolation:** A store surrounded by vacant businesses may struggle because of a lack of foot traffic.

- **Adjacency:** If a large industrial data center is built next to your house, your property value may take a hit.

All these factors have one thing in common: They require you to see, acknowledge, and question spatial locations, patterns, and distributions. Thinking spatially helps you ask better GIS questions and get more meaningful answers.

Getting better at spotting spatial patterns takes practice. Here are some tips to get you started:

- **Look at maps.** The more you study maps, the easier it is to spot patterns.

- **Notice how traffic flows.** Be aware of which roads get backed up at certain times of day.

- **Study aerial and satellite images.** These images give you a bird's-eye view of landscapes and how they change over time.

TIP

Seeing spatial patterns takes practice. You need to read maps, study satellite images, and most important, practice creating, querying, and analyzing spatial data with GIS. The more you do, the sooner you'll become a very spatial person. (See what I did there?)

Chapter 2
Learning GIS Core Concepts

IN THIS CHAPTER

» Finding out where spatial data comes from

» Familiarizing yourself with the basic types of maps

» Understanding how mapmakers flatten the Earth and measure it

Understanding the various aspects of GIS begins with a solid grasp of core concepts that lay the groundwork for your mapping and spatial analysis. In this chapter, I introduce you to the basic types of spatial data that are essential to any GIS project. You find out about different types of maps — reference, thematic, and topographic — and their unique purpose in visualizing spatial information. I also take you on a brief journey into the world of map projections, unraveling the methods mapmakers use to measure and flatten the Earth's surface.

Knowing these core GIS concepts is important not only for creating meaningful maps but also for making informed decisions that depend on spatial relationships. By mastering these fundamental aspects of GIS, you can build a strong foundation to begin interpreting, analyzing, and applying spatial data in a wide range of applications, from urban planning to emergency management and beyond.

Digging into the Foundations of Spatial Data

In GIS, everything starts with *spatial data*, the information that represents the location, shape, and attributes of features on the Earth. Understanding the types of spatial data and how to get these data into your GIS is fundamental to all your GIS work, from creating maps to performing complex spatial analyses.

Understanding spatial data types

In GIS, you use two basic types of spatial data, *vector* and *raster*, which I describe in this list:

- **Vector data are simply points, lines, and polygons.** These features represent the most common elements you use when drawing a map. For example, if you were to draw a map of your neighborhood, you would most likely use points to mark where the bus stops are located, lines to mark the roads, and polygons to mark the parks.

    In GIS, each vector feature contains spatial attributes so that your GIS application can position it in your map. But spatial attributes aren't all that vector data can contain. They can also include a table of attributes with descriptive information about each record. If you use GIS to create a map of your neighborhood, the polygon you create for the park boundary can also include the name of the park, whether it has picnic tables, and who's responsible for maintaining the grounds.

- **Raster data are defined by a grid of cells or pixels.** The easiest way to understand the raster data type is to think of a photograph. A photograph is like a mosaic made up of tiny colored tiles called *pixels*. When you take a selfie with your smartphone, its camera sensor captures an image as a grid of pixels, filling each pixel with a color that matches what you see. When you look at the image, you don't see all the little squares, or pixels, because your eyes blend them together into one smooth image of your smiling face (and that photo-bomber behind you).

    Similarly to vector data types, raster data also include spatial attributes so that your GIS knows where to position the raster data layer on your map. Unlike vector data, though, raster data don't have an attribute table associated with them. Instead, each raster cell contains a value such as elevation, temperature, or land cover. You may get raster data for your neighborhood in the form of satellite imagery. You can then use this imagery in your GIS to determine the amount of green space compared to streets and developments (houses and other buildings such as schools and gas stations).

I go into a lot more detail about vectors, rasters, and other spatial data types in Chapter 5. But for now, you can get a handle on the many uses of GIS by understanding vector and raster data types.

Acquiring your spatial data

After gaining an understanding of the basic types of spatial data, your next question may be where to get this data to put into your GIS. Both vector and raster data come from a variety of sources, including field surveys, satellite imagery, and public databases. Knowing where your spatial data come from is important for assessing its accuracy, relevance, and applicability to your GIS project.

You're probably already familiar with the use of GPS to collect a variety of vector data. Surveyors, engineers, geologists, and other professionals collect data using sophisticated GPS receivers that provide high accuracy and precision, which is critical for building roads, bridges, and other structures.

Data collected from smartphones and small, handheld GPS devices tend to be less spatially accurate, meaning that the mapped location probably doesn't exactly line up with where that point or line is on the actual surface of the Earth. However, the advantage of using these devices is their ability to quickly collect a lot of useful information about a specific area that you can then use to record and analyze what's going on there. For instance, you can use your smartphone's built-in GPS receiver to mark the location where you spotted that acorn woodpecker or to track your last hike through the forest. Many organizations use custom smartphone apps and handheld GPS devices to collect and analyze data on asset inventory, water quality sampling, and crop monitoring, to name just a few examples.

People use satellites, airplanes, and drones to collect raster data. You can get satellite imagery from a variety of online sources like NASA's Landsat program or Google Earth, or you can purchase custom data from commercial providers. With the proliferation of the small, uncrewed aerial system (sUAS) — commonly referred to as a drone — raster data is easier than ever to collect and use in your GIS projects.

Many public and nonprofit agencies provide geospatial data for free through open data portals. For example, the U.S. Census Bureau provides geographic boundary data that you can easily link to their demographic data. Another example is the OpenStreetMap Foundation, a crowd-sourced initiative providing free worldwide geographic data. Chapter 21 provides ten GIS data sources to help you begin your spatial data — collection adventure.

Discovering how you use spatial data

Spatial data plays an important role in decision-making by providing a geographical context for informed choices. If you're in the market to buy a house, GIS can help. Using spatial data on property values, neighborhood characteristics, and proximity to other aspects of the area that you may care about, such as grocery stores and dog parks, allows you to find a suitable location for your new home.

Spatial data are also crucial for analyzing areas at risk for flooding, wildlife corridors near planned highways, and regional demographics. Imagine that you're a transportation planner. Mapping wildlife corridors along with proposed highway alignments enables you to make informed decisions about which proposed alignment has the least impact on wildlife. Spatial data not only allow you to perform a variety of analyses but also give you a powerful tool for communicating your findings through the use of maps and graphical presentations.

Understanding Maps in GIS

Maps are fundamental to GIS. They provide a way for you to transform your spatial data into a meaningful, easy-to-understand visualization. By knowing how maps represent spatial data and the different types of maps you can create — such as reference, thematic, and topographic maps — you can learn how to visualize geographic relationships and patterns. In this section, I guide you through the essentials of mapping in GIS, including how maps use symbols to communicate information and the importance of scale for representing the Earth's surface.

Seeing how maps represent spatial data

GIS enables you to visualize your spatial data by using symbols that represent real-world features. It helps you discover patterns and relationships within your data so that you can make informed decisions. With the aid of colors, symbols, and lines to display attributes of your spatial data in a map, you can make complex information understandable to any audience. For example, a traffic map might use different colors to indicate congestion levels, simplifying the understanding of urban mobility issues. This capability to represent and analyze data visually is what makes GIS an invaluable tool for anyone needing to interpret geographical contexts quickly and effectively.

Reviewing types of maps

Anyone who creates a map must consider the content and subject of the map. They must assign a topic, or *theme*, to the map and set the individual details in the proper context. If you're creating a map to your house-warming party, for example, you must include roads so that your guests know how to get there. Landmarks and buildings may also help your guests get oriented to your part of town. You can even include text on the map, such as "Turn left at the ugly purple house."

REMEMBER

In a technical conversation, words often have different or more precise meanings than in a general conversation. The context and the theme are just as important in mapping as they are in a general conversation. A map needs a unique context, a specific vocabulary, and a set of rules that pull all these parts together so that both the mapmaker and the map reader can understand each other.

The easiest way to categorize maps is to separate them into three basic groups: reference maps; thematic maps; and topographic maps.

Reference maps

Reference maps offer a great deal of information in a single map. You commonly find these maps in atlases, which provide a collection of related maps all in one convenient location. Reference maps often cover very large portions of the Earth, which means they're created at small scales. Therefore, reference maps aren't designed to provide extremely accurate depictions of locations. Instead, they offer general patterns and relative positions of major geographic features.

Reference maps also commonly combine several types of information onto a single map. A typical reference map contains features such as political boundaries, transportation networks, and significant natural landmarks like rivers, lakes, and mountain ranges. By combining all these features into one map, you gain a more comprehensive view of how a region is structured and connected, both physically and administratively. Figure 2-1 shows a reference map of Morocco from the CIA World Fact Book, which includes political boundaries, roads, key cities, and prominent terrain.

Thematic maps

A thematic map provides detailed and precise information on a specific subject, as compared to a reference map (discussed in the preceding section), which provides a broad overview of various geographic features.

When you view a reference map, such as the one shown in Figure 2-1, you can see several data themes — in this case, boundaries, roads, cities, and terrain. The reference map doesn't present these themes in much detail, and it doesn't have the primary purpose of communicating specifics. For example, although the map includes transportation information, that information isn't complete; it's very general because of the small scale of the map. Also, the different types of routes aren't clearly identified. This map wouldn't help you much if you were trying to navigate within Morocco.

CHAPTER 2 **Learning GIS Core Concepts**   23

FIGURE 2-1: Reference maps may contain several types of information.

CIA World Factbook, World and Regional Maps, Morrocco/https://www.cia.gov/the-world-factbook/static/3745aa3d8e030c85cbd47bd5cf6fb5e7/morocco_map.jpg, last accessed on 26 March 2025/ Central Intelligence Agency (CIA)/Public Domain

For navigating, you need a *road map*. A road map is one example of a thematic map because it focuses on communicating information about roads. Although such a thematic map may include a few regional subdivisions and other landmarks, the majority of the map is designed to let you know which roads consist of two lanes and which of one lane; which are dirt roads and which are superhighways; which are one-way streets and which are two-way streets, and so on.

REMEMBER

Thematic maps are the most common maps that you'll use in your GIS activities. Although you may find reference maps useful for some projects that involve large study areas, the small scale of reference maps usually limits their usefulness for analysis. The more familiar you become with the wide variety of thematic maps available, the better equipped you'll be to determine the ones that best fit your GIS projects.

The list of potential thematic maps is staggering; here are a few examples:

>> **Medical maps:** Cancer mortality, spread of epidemics, and health hazards can be map subjects. Medical professionals may use such maps to determine characteristics of locations where some diseases occur, concentrate, and

24 PART 1 **GIS: Geography on Steroids**

spread. Knowing these factors can help professionals progress in their goals to prevent or cure disease.

>> **Hazardous material maps:** These maps give transportation companies information about where trucks carrying hazardous material can travel, as well as provide the relative locations of people at risk (for example, the elderly or hospital patients).

>> **Law enforcement maps:** These kinds of maps show the police crime hot spots, locations of officers, crimes by time of day and season, and socioeconomic factors that may affect crime rates.

>> **Utility maps:** Gas and electric utility infrastructure maps can include data about the type of equipment, its condition and age, and many other factors affecting the energy companies' ability to provide and sustain services.

If a feature appears or an event occurs on the surface of the Earth, it can be — and often has been — mapped.

Topographic Maps

Topographic maps depict both natural and human-made features of a region in detail. The most common characteristic found on topographic maps are elevation *contour lines*, although these maps may also include features such as roads, urban infrastructure, rivers, and other landmarks. Figure 2-2 shows a topographic map of a portion of Grand Canyon National Park in Arizona. Each contour line represents a specific elevation. Note how the lines are spaced at fixed intervals to depict the shape of the Earth's surface.

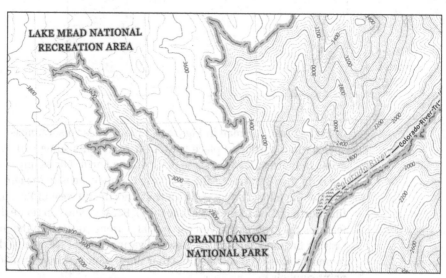

FIGURE 2-2: A portion of a topographic map showing an area in Grand Canyon National Park.

U.S. Geographical Survey / https://www.usgs.gov/faqs/are-usgs-topographic-maps-copyrighted, last accessed on 26 March 2025 / Public Domain.

CHAPTER 2 Learning GIS Core Concepts

Topographic maps help you understand terrain and navigate effectively. They are most commonly used in planning, surveying, hiking, conservation, and military operations.

Interpreting symbols

Knowing how symbols on a map communicate information can enhance your understanding of the different types of maps and their uses. Lines, shapes, colors, and words comprise the language of a map, providing a visual shorthand for the geographic items they represent.

A map may contain various shaded lines, little white squares, circles, green patches, words, letters, and numbers. You may find the number of symbols a little overwhelming. As a GIS and map user, you need to recognize that all the symbols have a purpose. They don't just happen (much to the chagrin of the people who have to make them). Cartographers plan the selection, size, placement, and design of all these symbols before putting them on the map. Figure 2-3 shows a road map, representing part of England, that contains various symbols.

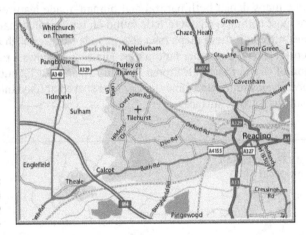

FIGURE 2-3: Symbols on a road map of England.

At first glance, the map in Figure 2-3 may appear disorganized, but each symbol appears in a particular place because it represents some real-world feature that exists in more or less the same place on the Earth. For example, the city of Reading is on the right side of the map, and Pangbourne is to the left and up a bit. This positioning tells you that Pangbourne is north and west of Reading (if you assume that north is toward the top of the map). Because you can relate the symbols to the features they represent, you can quickly orient yourself to the map. Without ever visiting England, you can see how the various towns are arranged in space relative to one another. You can also see that lines, which represent roads, connect these point symbols.

26   PART 1 **GIS: Geography on Steroids**

Now look back at Figure 2-2, the topographic map showing a portion of Grand Canyon National Park. Note how the use of contour lines gives the illusion of elevation change. Cartographers have been using this technique for more than 200 years.

REMEMBER

The symbols you use on your maps will vary depending on the purpose and style of the map. Many maps follow standardized symbols and color palettes to help make them universally understandable. For example, the U.S. Geological Survey (USGS) uses standard map symbols on all their map products. These maps are updated periodically, using the standardized symbols to maintain consistency with previous versions of the maps. It's always a good idea to check with the organization you're making a map for to see whether they have any standardized map symbols or color palettes for you to use in your map.

Chapter 3 explains how to translate map symbols.

Getting the scale right

REMEMBER

When compared to thematic maps, reference maps are often created on too small a scale to be particularly useful for GIS. But even thematic maps come in many different scales. A good rule of thumb is the larger the map scale, the smaller the area covered and the greater the detail. Larger-scale maps are generally better for your GIS activities because they provide the largest amount of detail.

If all maps were of the same highly detailed scale, you wouldn't have to think about what scale a particular map uses. Unfortunately, you find many maps at many different scales. In fact, most GIS databases contain a lot of data from differently scaled maps. Each map has more or less detail, more or less positional accuracy, and more or less informational content.

Naturally, the portion of the Earth represented on a map is much larger than the drawing of it on your map. The map and the Earth are at different scales. *Scale* determines how much or how little detail your map can hold.

You can often find a map's scale represented by a graphic bar and a fraction that shows the relationship between the size of the map in the *numerator* (the fraction's top part) and the size of the Earth in the *denominator* (the fraction's bottom part). Using this mathematical approach, the smaller the fraction (one with a small numerator and a large denominator), the smaller the scale. The smaller the map scale, the larger the amount of the Earth that map represents.

To get this representation to work, the mapmaker (*cartographer*) uses a little trick or two. Instead of including all the detail of the Earth, the mapmaker selects which objects to show and decides how much graphic detail of each object they can omit

CHAPTER 2 **Learning GIS Core Concepts** 27

to save space while still expressing the necessary information. Table 2-1, from the U.S. Geological Survey of typical map scales, shows the amount of the Earth that different map scales represent.

TABLE 2-1

Map Scales

Scale	1 inch represents approximately	1 cm represents
1/24,000	2,000 ft. (exact)	240 m
1/50,000	4,166 ft.	500 m
1/62,500	1 mile	625 m
1/63,360	1 mile (exact)	633.6 m
1/125,000	2 miles	1.25 km
1/250,000	4 miles	2-5 km
1/500,000	8 miles	5 km
1/1,000,000	16 miles	10 km

USING A WIDE RANGE OF DATA

Thematic maps come in all sorts of sizes and shapes, and they can cover many kinds of topics. When you think about all the activities that actually occur on the Earth, you start getting some idea of the number of potential thematic maps. Here's a short list of some of the kinds of maps that contain socioeconomic information and how you may want to recognize and apply these patterns:

- **Median household income:** If you're trying to start a business that sells expensive jewelry, you probably want to place your store in an area that has a high median household income.

- **Vehicles available:** If you're a city planner expanding public transit, you want to prioritize areas with the greatest need, which typically is where fewer vehicles are available per household.

- **Mean number of children under five years old:** If you want to sell baby clothes, you ideally want to establish your store close to large concentrations of people who have young children.

This set of examples uses a useful concept in geography: All things are related to each other in geographic space, but close things are more related than far things.

Both the mapmaker and the map user need to understand map scale so that they can (respectively) create and read the map properly.

Flattening the Earth

Except for members of the Flat Earth Society, everyone accepts that the Earth is roughly spherical. This spherical shape has some major drawbacks for the mapmaker faced with producing a flat map that correctly represents the shapes, angles, distances, and sizes of objects on the Earth. In simple terms, projecting the Earth's surface onto a plane will always result in some sort of distortion. Peel an orange and try to flatten the peels on a table; now you have a visual idea of the problem that the mapmaker faces. Visualizing the projection of the three-dimensional Earth onto a flat surface using an orange peel is simple. Making a map representing the three-dimensional Earth is much more involved. But have no fear: In this section, I walk you through the process.

Measuring the Earth

Way back in the third century B.C., a Greek scholar by the name of Eratosthenes first calculated the circumference of the Earth. Often considered the founder of geography, this ancient Greek scholar was the first to record the use of a grid system to map, or model, the Earth's surface. How the Earth is measured and modeled for mapping purposes is the basic concept behind geodesy. *Geodesy* is the science behind measuring and understanding the Earth's shape, size, gravity field, and how those aspects change over time.

In the early days, geodesy involved measuring the Earth using techniques like *triangulation,* which determines distances by measuring angles between points to form a series of interconnected triangles. This method worked fairly well. In fact, it's the method Eratosthenes used to measure the circumference of the Earth, and he came within two percent of the actual value. That's pretty close, if you ask me! Today, we have advanced technologies, like GPS, that use satellites to continuously monitor the size and shape of the Earth.

REMEMBER

Identifying, locating, and measuring features on the Earth's surface for use in mapping, engineering design, construction, and other related activities require more than what you might learn by simply peeling an orange. To start, you need a basic understanding of some key concepts for representing the Earth's three-dimensional, curved shape.

CHAPTER 2 Learning GIS Core Concepts   29

The Earth is not a perfect sphere. It's flattened at the poles, it bulges at the equator, and it's lumpy with mountains and canyons across its surface. These imperfections cause gravity to pull on everything in different directions and at different strengths. As you may imagine, these complexities make it difficult to map Earth's surface. Fortunately, geodesists, or scientists who specialize in geodesy, have developed two mathematical models of the Earth's surface, the geoid and the ellipsoid, that make it easier to measure and map the Earth, despite its many imperfections. Although they may sound similar, the geoid and the ellipsoid are used differently.

The *geoid* is a model of Earth's shape that accounts for its uneven gravity field. Rather than being a perfect sphere, the geoid reflects the pull of Earth's gravity, with areas that rise and fall due to the variations in mass across the planet. The geoid is important for defining accurate elevations relative to the average sea level, but because it's so complicated, it's not ideal for mapping or surveying.

This is where an ellipsoid comes in. An ellipsoid is a mathematically simplified shape that approximates the Earth's surface, smoothing out the lumpiness to create a more manageable surface than a geoid provides. An ellipsoid is the shape commonly used for building reference systems, which are necessary for accurate mapping, surveying, and navigation. A *reference system* is a framework used to define and measure locations on the Earth's surface.

I have thrown out a lot of new terms in this section, so I know it can be tricky to take them all in. But don't fret! Even seasoned GIS professionals often confuse the terms *geoid* and *ellipsoid*, so here's a more in-depth description of each:

- » **Geoid:** The geoid is a model of the Earth's shape based on gravity. It represents an imaginary surface where the ocean would settle if no tides, winds, or currents existed. Essentially, it would be as though the average sea level extended across the entire planet. This surface is shaped by variations in Earth's gravity, meaning that it's not a perfect sphere or ellipsoid. Scientists use the geoid as a reference to measure elevation and understand how gravity affects the planet.

- » **Ellipsoid:** An *ellipsoid* is a sphere shape that is slightly flatter at the poles and thicker at the equator. It serves as a reference surface for measuring the Earth. Before Earth-orbiting satellites came along, people created ellipsoidal models for a best fit across continent-sized regions. But an ellipsoid that fits North America may not be useful if you're working in Australia. As a result, many different ellipsoids have been created to get accurate measurements based on where your project is located. Today, the most commonly used ellipsoids are the Geodetic Reference System of 1980 (GRS80) and the World Geodetic System of 1984 (WGS84).

Figure 2-4 is a simple depiction of how the geoid and ellipsoid relate to each other and the Earth's surface.

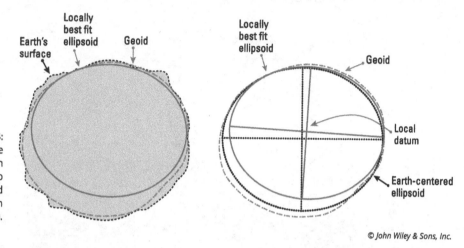

FIGURE 2-4: A model of the Earth (left) and an ellipsoid fitted to the geoid based on location (right).

© John Wiley & Sons, Inc.

REMEMBER

A reference system has three main components: an ellipsoid, a datum, and a geographic coordinate reference system.

Ellipsoids are the foundational part of a reference system. They help build the framework for locating and mapping both natural features, like mountains and rivers, and human-made features, like buildings and roads, in a GIS.

An ellipsoid alone is not enough to locate and map features on the Earth. To make accurate measurements and locations, your ellipsoid needs to be referenced with respect to the Earth itself. Enter the datum. A *datum* provides the instructions for where that ellipsoid is located and lined up with the Earth. Here's an example: Imagine that you have a globe in your hands. It represents an ellipsoid, or a smooth, simplified version of the Earth. Take that globe and set it on a table; then orient it so that everything lines up correctly with how the Earth is oriented based on where you're standing. The datum would be where you decided to place the globe and angle it so that it lines up with Earth. Different datums might put the globe in a different location on the table or at a different orientation. It all depends on what part of the Earth you want to match best.

Now that you have your globe (ellipsoid) and its location (datum) figured out, you need a method for locating things on it. For this task, you bring in a *geographic coordinate reference system (CRS)* of latitude and longitude, as shown in Figure 2-5. Lines of latitude, also called *parallels*, run east and west. The equator defines the midway point between the poles and defines the origin or zero latitude. Lines of longitude, also called *meridians*, run north and south. The zero longitude is known as the prime meridian. These lines of latitude and longitude encompass the Earth, forming a grid network.

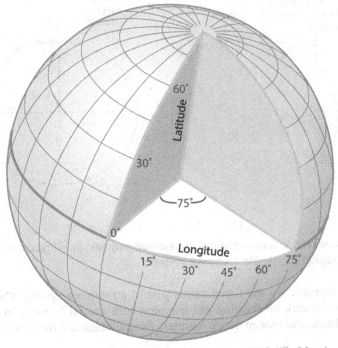

FIGURE 2-5: Earth's geographic coordinate reference system.

© John Wiley & Sons, Inc.

The ellipsoid, datum, and geographic CRS are what define the three-dimensional model of the Earth. Devices that use GPS technology use this three-dimensional model to determine where things are on the Earth. This model is how your smartphone knows how and where to mark the spot where you saw that acorn woodpecker on your last hike.

Understanding map projections

Map *projections* — the outcome of converting the spherical Earth to a flat surface — come in many different types, from contiguous to interrupted, and from those that look like photographs of the Earth to those placed on cones or cylinders. Because we often use flat maps for navigation, planning, and analysis, converting the three-dimensional Earth to a two-dimensional surface is essential. Without projections, large-scale paper maps and digital maps wouldn't be possible.

Cartographers often refer to the three families of map projections, shown in Figure 2-6, as cylindrical, conic, and planar (also called *azimuthal*). These projections describe the surfaces onto which the surface of the Earth would be projected (hence the term *projection*) if a light bulb could be put in the center of the Earth.

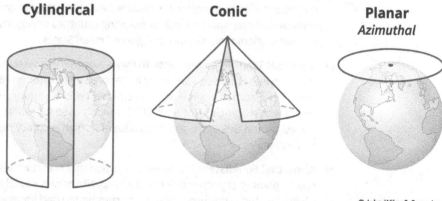

FIGURE 2-6: The three families of map projections.

© John Wiley & Sons, Inc.

A geodesist once told me, "Nothing is certain except death, taxes, and map distortion." The curved surfaces of the Earth simply don't cooperate when you try to make them flat. Map projections distort area, distance, shape, and direction. No map projection can preserve all four of these properties simultaneously. This certainty — that distortion is unavoidable — makes choosing the right map projection essential for ensuring accurate measurements and minimizing distortions that could impact the map's intended purpose. Using the wrong projection for a project can lead to serious errors. Imagine planning fuel for a transatlantic flight and underestimating the distance. That might not end well!

TIP

When working with GIS, pick the map projection that best represents the properties you want preserved, minimizing the distortion in the areas that matter most for your work. For example, if your analysis involves changes in area, a map projection that preserves area is the best type to use to display your results.

Hundreds of documented map projections exist, with many custom projections created by cartographers and geodesists to meet specific mapping needs. Some common map projections include

» **Web Mercator** is a cylindrical projection that preserves local shapes and angles. Originally developed for navigation, the Mercator projection was great for seafaring but not much else because it significantly distorts area, especially near the poles, making regions like Greenland look much larger than they

CHAPTER 2 **Learning GIS Core Concepts** 33

really are. Today, Web Mercator is the standard for most online maps because it provides a consistent, familiar look and allows for seamless zooming and panning.

» **Lambert Conformal Conic** is a conic projection that preserves shape and is most accurate along lines of latitude (standard parallels). Developed for mapping mid-latitude regions, it became (and still is) the top choice for aeronautical maps and is ideal for mapping countries or regions with an east-west orientation, like the contiguous United States.

» **Universal Transverse Mercator (UTM)** is a coordinate system that divides the Earth into 60 zones, each using a Transverse Mercator projection to minimize distortion in small areas — typically regions smaller than a country that fit within a single UTM zone. UTM is widely used for topographic maps, surveying, and GPS applications because of its high accuracy over local and regional scales.

» **Azimuthal Equidistant** is a planar projection that preserves distances from a central point to any other point on the map. Historically useful for air and sea navigation, today this projection is also commonly used for maps showing distances from a specific location, such as depicting travel routes or communication networks radiating out from a city to other points around the world.

Using coordinate systems

No matter what map type, projection, or scale you use, you need to find your way around on your maps — and that's what coordinate systems help you do. On a globe, you use the latitude-longitude system. On flat maps, you have to convert this geographic grid to something a bit more practical. Ultimately, all coordinate systems are based to some extent on the geographic grid.

But after you take a spherical object such as the Earth and flatten it, you need to have a grid that conforms to this new surface. In other words, after you project the spherical Earth onto a flat surface, you need to put a set of direction lines on that flat surface that conform to the distortions the projection just produced. So coordinate systems are often linked to specific types of map projections.

Coordinate systems fall into two types:

» **Geographic (refer to Figure 2-5):** Uses a three-dimensional spherical surface to determine locations using latitude and longitude and measured in degrees. You use this system mostly for global applications when you need to account for the curvature of the Earth.

» Projected: Translates the spherical coordinate from a geographic coordinate system onto a flat plane, using map projections to minimize distortion. A projected coordinate system is a reference framework that uses a specific map projection to define how coordinates (such as x and y values) are plotted on a flat surface. One example of a projected coordinate system is UTM (described in the previous section), which is based on the Transverse Mercator projection. Unlike the Web Mercator, which is widely used for navigation and web mapping, UTM divides the world into zones, each with its own coordinate system, minimizing distortion for regional-scale mapping. It also specifies zones, scale factors, and the WGS84 datum to provide a standard way of mapping locations globally.

Defining land boundaries and ownership

In addition to geographic and projected coordinate systems, specialized coordinate systems define land ownership, manage resources, and establish property boundaries.

The U.S. Public Land Survey System (PLSS) was established in the late 1700s to divide the ever-increasing lands that the U.S. government acquired. The PLSS is similar to a rectangular coordinate system, but it's not directly linked to any particular map projection.

Unlike coordinate systems that use projections or grid coordinates to define a point's location on the Earth's surface, the PLSS uses a grid of townships and ranges to describe areas of land. This makes it ideal for legal descriptions, land sales, and the management of natural resources.

The notation that identifies a parcel of land starts with the smallest subdivision of land, followed by the section number and then township and range numbers. A legal land description based on this system includes the state name, principal meridian name, township and range designations, and section number.

Such land subdivision systems enable you to build GIS databases that have land ownership correlated to other factors, such as zoning restrictions or price per acre.

MAPS AS QUESTIONS AS WELL AS ANSWERS

If you're like most people, you view maps as answers to the question, "Where is _____?" — filling in the blank with whatever you're looking for. Although maps do display locations and distributions of geographic features, for GIS professionals, those functions are just the beginning. A map not only shows where things are but also raises more questions about the patterns and relationships among the many features on your map.

For example, if you're a crime analyst with a map of auto thefts for your city, you might start by asking, "Where are auto thefts occurring?" But you want to know more, such as the following:

- Do auto thefts occur more in some areas than in others? Does your city have auto theft "hot spots"?
- Do auto thefts increase at certain times of the day or year (does a temporal pattern exist)?
- Do other distributions seem to match that of the auto thefts? Or, stated differently, does a spatial correlation exist between auto thefts and other factors, such as neighborhood type, access to major highways, or gang activity?

Thinking about your map in this way transforms it into both an *answer* and a *question*. As you explore data with GIS, you begin asking "What explains this pattern?" GIS gives you tools to answer that question. If you can explain the pattern, you can begin to predict future trends and make informed decisions.

For example, if you're a transportation planner, you may notice increased traffic in a particular neighborhood. Using GIS, you discover that the increase in traffic is related to population growth. Armed with that information, you can recommend future improvements to the transportation system. GIS tools enable you to analyze patterns, predict future changes, and make data-driven decisions. In a way, GIS makes you a really great prognosticator.

> **IN THIS CHAPTER**
>
> » Translating map symbols
>
> » Spotting patterns
>
> » Analyzing the patterns you identify
>
> » Applying your conclusions to the decision-making process

Chapter 3
Reading, Analyzing, and Interpreting Maps

Maps are more than just static images; they're stories waiting to be uncovered. Exploring GIS opens a world of content-specific, or *thematic*, maps, each designed to reveal specific aspects of geography or human activity. Understanding these maps and their symbols is key to unlocking their stories. This knowledge helps you figure out the best way to analyze your data and make informed decisions.

As you become familiar with different map types, especially thematic ones, you start noticing patterns in the symbols. Spotting these patterns can be tricky at times. But with practice, those symbols and patterns become the language of landscapes, cities, weather, and more.

In this chapter, you find out how to identify patterns in map symbols. You also discover how to ask the right questions, use GIS analysis to answer them, and make informed decisions based on what you discover.

Making Sense of Symbols

Many types of maps exist, each with its own unique set of symbols. *Symbols* can be lines, objects, or pictures that represent real things on the ground. With so many different map types and symbols, interpreting them can be confusing. But have no fear! In the next sections, I show you how to make it easier on yourself by focusing on the dimensions and measurement scales used in mapping and GIS.

Categorizing the space on a map

When cartographers talk about "space on a map," they're not talking about that spot on the map where early cartographers added elaborate drawings of sea monsters and mermaids. On the contrary, they're referring to the different ways that geographic features are represented using symbols. Understanding how cartographers categorize space makes it easier to interpret a map. The types of space you find on maps include the following, with each depicting a different dimension of the real world:

- » **Points:** Considered zero-dimensional because they don't take up space, points represent specific locations like towns, wells, or schools. Maps typically use symbols such as dots or small icons for point features.
- » **Lines:** One-dimensional and measured by length, lines represent features like roads, rivers, or railroad tracks. Different line styles and colors on a map can symbolize different types of roads or waterways.
- » **Areas:** Areas are two-dimensional, measured by both length and width. Areas represent regions like parks or lakes, often using colors or patterns to designate their location on the map.
- » **Volumes:** Representing three-dimensional space that is measured by length, width, and depth or height, volumes indicate features like mountains. These features are often simplified as area symbols on a map. Understanding volume spaces is important when analyzing features such as groundwater reserves, oil and gas deposits, or the movement of magma beneath volcanoes.

When it comes to map symbols, cartographers treat volumes and surfaces pretty much the same. In contrast to volumes, however, think of surfaces as "blanketing" something rather than including what's inside. For example, if you're interested in the vegetation covering a mountain, you focus only on its surface. But if you're interested in the amount of coal underground, you need to know not just the surface of the ore seam but also its entire volume, because extraction requires knowledge of the full depth and extent of the coal deposit.

REMEMBER

Maps symbolize four types of spatial objects — points (zero-dimensional), lines (one-dimensional), areas (two-dimensional), and volumes (three-dimensional) — with each representing different aspects of space on the map.

Understanding levels of measurement

You measure all map data, whether they are points, lines, areas, and volumes, in one of four scales:

» **Nominal:** This scale works at the most basic level, identifying objects by their name — a school, a house, a highway, or a parcel of land. *Nominal* measurements are just labels, so you can't really compare them to each other; it's like comparing apples and oranges.

» **Ordinal:** Ranking things by a general category is using an *ordinal* scale. For example, you might categorize houses as small, medium, or large. This scale gives you a sense of order but not specific differences between the categories.

» **Interval:** This scale adds more detail, allowing you to measure the exact differences between things. For example, you can say that 10 is exactly 90 units less than 100. This scale not only shows order but also provides measurable *intervals* — the equal differences between values, unlike ordinal scales that lack fixed differences.

TIP

Because points are technically zero-dimensional, measuring them at the interval scale may seem a bit, well, pointless. But points can still represent something that's measurable, like the average soil temperature at a specific spot.

WARNING

You can't make ratios out of interval measurements. For example, if the soil temperature is 32 degrees Fahrenheit in one location (A) and 16 degrees at another (B), you can say that A is 16 degrees warmer than B. But you can't say that A is twice as warm as B because Fahrenheit doesn't have a true zero starting point. The same goes for dates on calendars: Different calendars have different starting points (for example, the Mayan versus the Gregorian calendar), so you can't use them to calculate absolute ratios.

» **Ratio:** Items that have an absolute starting point, such as population (where zero means none), are measured on a ratio scale. It's called *ratio* because you can compare these numbers using ratios. For example, a city (A) with 2,000,000 people has twice the population of another city (B) that has 1,000,000 people — a ratio of 2 to 1 or 2:1. Other examples of ratio data include the size of a land parcel, the length of a border, or the volume of a buried ore deposit.

Here's a bite-sized reference for the four primary scales of geographic data measurement:

Type of Scale	How It Works
Nominal	Categorizes objects using names or labels, without any inherent order
Ordinal	Ranks objects in order, but differences between ranks are not measurable
Interval	Measures differences using equal increments, but with an arbitrary starting point
Ratio	Measures differences using equal increments and with an absolute (true zero) starting point, allowing for ratio comparisons

Matching symbols with data types

Figure 3-1 shows how the different measurement scales relate to symbols used for geographic features on a map. The table shown in the figure includes examples of the type of geographic features you might run across in your GIS work.

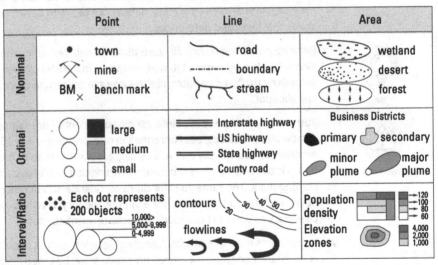

FIGURE 3-1: Geographic features and how they're measured.

© John Wiley & Sons, Inc.

In Figure 3-1, you see symbols for points (zero-dimensional), lines (one-dimensional), and areas (two-dimensional). Also note that the list of measurement scales combines interval and ratio scales because they share the same types of symbols.

40 PART 1 GIS: Geography on Steroids

Working with nominal features

Figure 3-1 displays nominal data in the first row, showing typical geographic objects like a town, a mine, a stream, and a forest. Each feature is represented by a unique symbol on the map. These features are *noncomparative* because you can't compare them to each other. You can't easily compare a town like Oakmont to the Black Hills mine; they're simply different kinds of features.

Other noncomparative features include

- » **Line features:** Streets, rail lines, and boundaries are unique entities and can't be compared to each other. Geographers say that they're not the same *kind* of feature.
- » **Areas:** A wetland, a wildlife range, government-owned land, or zoning types are all unique and noncomparable.
- » **Volume features:** Water aquifers, hills, and ore bodies occupy space, or volume, and are named on maps, but you can't directly compare them.

Depicting ordinal features

You can add *ordinal* (ranked) attributes to points, lines, areas, and volumes. You can rank inhabited regions as hamlets, villages, towns, and cities to show their relative sizes. This ranking indicates that the features aren't only of the same kind — inhabited places — but are also of different sizes of inhabited places.

Geographers call these distinctions *comparisons of kind* and apply them like this:

- » **For lines:** Rank roads from unpaved dirt roads to interstate highways to indicate their relative sizes (ordinal).
- » **For volumes:** Categorize the ore quality as low-, medium-, or high-grade to show differences.

Comparing data using interval and ratio scales

Although interval and ratio data use the same symbols on maps, mapmakers have many options for designing these symbols. One common method, known as *graduated symbols*, is to change the size of the symbols to match the size of the features they represent. Each symbol can be *graduated* (or sized) to the feature it represents. Easy squeezy, right?

Not so fast — there's a catch. Imagine you're mapping 3,000 counties in the United States. If each county has, say, 30 cities, that comes to 90,000 cities needing symbols! Even if you map only one city for every third county, you still have 1,000 symbols to size. Just imagine the frustration of someone who has to identify the differences among that many symbols!

The good news is that a simpler way exists to design symbols. Instead of giving each city a unique symbol size, you can group them into *classes* that represent a range of values. For example, cities can be grouped by population: 0 to 10,000; 10,001 to 100,000; 100,001 to 1 million; and more than 1 million. This approach makes it much easier for readers to understand your map.

Cartographers use this grouping method not just for points, but for lines, areas, and volumes, too. Here are some examples of applying symbology to different groupings:

- **For lines:** Use different line thicknesses to show river flow levels (see Figure 3-1).
- **For areas:** Represent the potential area of a hazardous spill using tear-drop-shaped symbols that vary in size.
- **For volumes:** Display the size of ore bodies with irregularly shaped area symbols that increase in size or give a 3D effect.

REMEMBER

Cartographers spend a lot of time and effort creating standardized symbols and class ranges to make maps easier to read. The good news is that they've already done this work for you. The so-so news is that grouping map data into classes can be less precise than the original data. Still, grouping helps make complex information much easier to understand.

Recognizing Patterns

One cool thing about maps is that their symbols represent a scaled-down version of real-world geography. Cartographers arrange the features, symbols, and background elements to closely resemble the actual objects on the ground. This approach is important because patterns of geographic features often relate to an underlying process, revealing cause and effect.

That underlying process usually started in the past, continues in the present, and is likely to happen in the future. Although this continuity isn't always the rule — because processes can speed up, slow down, or even stop — recognizing patterns helps you describe, explain, and even predict what might happen next. But first, you need to spot these patterns!

When you identify patterns, you're looking for predictability in how the objects are arranged. For example, if you map trees that are dying from a disease, you can predict that the nearby trees are likely to become infected as well.

TIP

Each person interprets patterns differently, depending on their experience and background. The trick to using GIS effectively is to spot patterns that might not be obvious, like clusters of trees, houses, roads, rivers, or any other feature you encounter.

Identifying random patterns

Patterns are based on how close together or far apart objects are from each other. Some objects, like dandelions in your yard, appear at random because of the way they spread (like in Figure 3-2). Dandelion seeds are scattered by the wind, animals, or even people, leading to varied distances between the individual plants. Some dandelions are close to each other, some are far apart, and others grow somewhere in between, producing a random distribution pattern.

FIGURE 3-2: Dandelions create a random pattern.

Samir Behlic/Wirestock/Adobe Stock Photos

Finding clustered patterns

Some features naturally *cluster* together. For example, auto thefts often form *hot spots* in specific neighborhoods (as shown in Figure 3-3). These hot spots reveal underlying factors, such as the availability of high-value vehicles or easy escape routes. You might find auto thefts concentrated in areas with less surveillance or near major highways. After you start spotting spatial patterns, you begin to connect them with possible causes and effects.

CHAPTER 3 **Reading, Analyzing, and Interpreting Maps** 43

WARNING

Be careful when interpreting patterns. For example, maps showing clusters of tornados may simply show where more people are available to report them and not necessarily show a weather pattern. It's important not to jump to the wrong conclusion, such as assuming that more people in an area means more tornadoes. The number of sightings is related to population, but not to the actual cause of tornadoes.

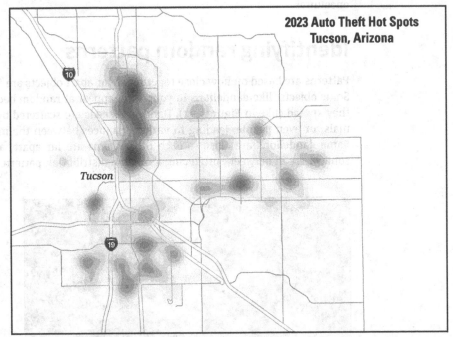

FIGURE 3-3: A clustered pattern appears in a map of crime hot spots.

© John Wiley & Sons, Inc.

Observing uniform patterns

Uniform patterns are the easiest to spot. An orchard, like the one in Figure 3-4, is a classic example of a uniform distribution. Each tree is spaced practically the same distance from each other. This type of pattern rarely occurs naturally. It's obvious that the pattern is driven by some form of human action — in this case, planting.

FIGURE 3-4: Humans create a uniform pattern in an orchard.

Robert Peak/Adobe Stock Photos

Seeing patterns among diverse features

In GIS, features sometimes show up in certain places but are missing in others. The way these features are arranged — whether by size, shape, or position — can reveal spatial relationships between different things on the map. Consider these examples:

- A map may show wetlands in specific areas, usually occurring where the landscape has a depression.
- Commercial development occurs only in select parts of a city, a pattern heavily influenced by city codes and zoning laws.
- A species shows up in one place and not another. Typically, plants and animals adapt to differences in the environment at various locations.

Each of these patterns reveals clues about the processes at work (cause and effect) in the same way that random, clustered, and uniform spacing do (as I discuss in the preceding sections). After you spot these patterns, you can explore their causes and use this knowledge to inform your analysis and decision-making.

Describing patterns with linear features

Linear features, like roads and rivers, show patterns of connectivity, flow, and linkages. You don't find these patterns in other feature types.

CHAPTER 3 **Reading, Analyzing, and Interpreting Maps** 45

Take a look at a U.S. road map. You likely immediately notice a pattern: Highly populated areas, like the eastern seaboard, have lots of highways. The map shows tons of connections offering a variety of ways to get from one place to another. Other areas, like the Great Plains, have many fewer highways, with limited connectivity. The number and arrangement of connections reveal patterns in the network.

Some road networks form closed loops, or *circuits*, just like electrical circuits do. Circuits in road networks allow traffic to flow around congested areas or obstacles. A common example is the loop surrounding major cities, known as a beltway, ring road, or outer loop, depending on where you're from. This type of circuit is designed to let drivers bypass the congested city center (see Figure 3-5). When you spot such a loop, it's often a sign of high population density in the area.

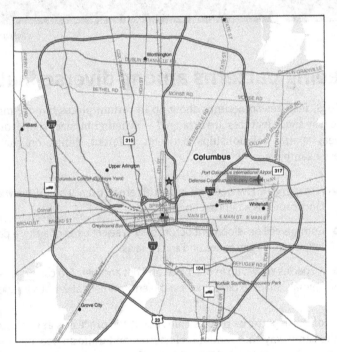

FIGURE 3-5:
A highway outer loop allows traffic to flow around the city.

Understanding the repeated sequence of shapes

You can easily spot patterns in a repeated sequence of shapes. The Earth's landscape — like that plaid shirt hanging in the back of your closet or a pair of argyle socks — is filled with repeated patterns. Flowing waters, for example, connect and flow in designs that create shapes, like branches of a tree or spokes on a wheel. It all depends on how the land shapes them. Figure 3-6 shows a thumbnail of each of the following patterns:

- » **Dendritic:** The easiest pattern (for me, at least) to remember is called *dendritic* and looks like the branches of a tree. These branches, called *tributaries*, go out in all directions and seem to have a mind of their own. The dendritic pattern usually forms in flat areas without strong rock formations, thereby allowing water to flow in seemingly random ways.

- » **Radial:** Another common pattern that flowing waters create is called *radial*. A radial pattern looks like a dendritic pattern, except that all the water flows outward, away from a center, like the spokes of a wheel. A radial pattern occurs when a hill or mountain at the central point works with gravity to cause water to flow down and out from the elevated center.

- » **Centripetal:** The opposite of a radial pattern, a *centripetal* pattern occurs when a low spot or depression affects the flow. The radial pattern is reversed, and the stream flows toward the central low spot, like a basin.

- » **Parallel:** *Parallel* and *sub-parallel* streams run along the gentle slopes that result from either natural topography or from human activities like road construction or mining.

- » **Trellis:** A *trellis* stream pattern results from rock layers, or *strata*, that have been exposed and folded over time, guiding streams along jointed and intersecting lines. This pattern features parallel main tributaries that follow folded rock layers, whereas perpendicular subsequent streams cut across weaker zones, creating right-angle junctions. The overall effect resembles a garden trellis or a city street grid.

- » **Rectangular:** A *rectangular* stream pattern features strong right-angle turns, often as the result of cross-cutting joints in the underlying rock. This pattern also occurs in some human-made neighborhoods where drainage ditches are strongly oriented along a grid.

- » **Annular:** An *annular* stream pattern occurs in areas with an eroded dome or basin. The erosion forms a series of circular rings with fractures at right angles to the rings, forcing water to flow in the interrupted portions along the circular paths. The term *annular* comes from the geometric form *annulus,* which is a ring.

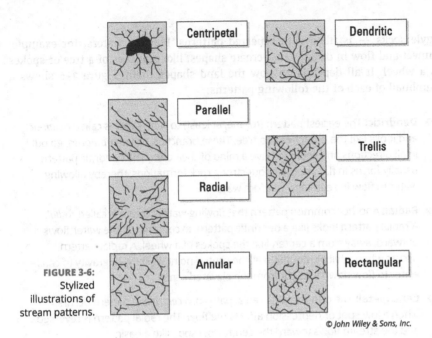

FIGURE 3-6: Stylized illustrations of stream patterns.

© John Wiley & Sons, Inc.

Analyzing and Quantifying Patterns

Recognizing patterns is a key skill for every GIS user, especially an analyst. Patterns occur for a reason, even if it's a random one. As a GIS user, you analyze patterns to determine whether they result from random, clustered, or uniform processes (see the "Recognizing Patterns" section, earlier in this chapter). After you understand the process behind the pattern, you predict future patterns. For example, if you notice that a portion of a crop isn't doing well, the soil might be low in nutrients — perhaps because of overplanting or underfertilization.

Knowing your geometry and patterns

Understanding the geometry and patterns in your maps helps you analyze them more effectively. Here are three basic approaches for working with patterns:

>> **Describe** the patterns accurately so that anyone can understand them.

>> **Compare and contrast** these patterns quantitatively with other features.

>> **Analyze** the quantities to assess their pattern or shape and create a baseline for tracking changes over time.

ANALYZING PATTERNS IN THE REAL WORLD

Recognizing, quantifying, and analyzing spatial patterns give you the power to interpret the meaning and importance of these patterns. This chapter includes examples about the significance of patterns, but the list of pattern types and causes is nearly limitless:

- Wildlife specialists study the locations, range, and movement of animals to understand what causes changes to their population and habitat.

- Urban planners study a variety of patterns including street networks, utility grids, population distribution, and traffic patterns. Studying multiple patterns allows urban planners to plan for effective growth that provides access to housing, jobs, and essential services like power, water, and sanitation.

- Military strategists analyze the locations and distributions of troops and equipment to effectively deploy their forces.

- Criminologists study distributions and concentrations of crime to help protect the public.

- Utility companies examine their distribution networks for wear or damage caused by weather and age.

- Businesses analyze the distribution and demographics of potential customers to determine where to open new locations.

- First responders need to know street connectivity and traffic patterns at different times of day.

GIS software automates much of this analysis by providing the values you need for future analytical techniques. It counts objects, determines the absolute locations of features (points, lines, and polygons), and measures objects and the spaces between them. These techniques are the same ones that you use to calculate geometry and orientation, which I cover in Chapter 10.

Using GIS for the analysis

When you load spatial data into your GIS, you specify where each feature is located and how it relates to the Earth's surface. Your GIS software then takes over, quickly calculating useful information like:

>> The number of points, lines, and polygons (areas) in the dataset.

» The size of individual polygons and the total area for loaded data layer.

» The total lengths of certain line types.

When it comes to describing patterns and spatial geometry, GIS does most of the heavy lifting for you, but not all of it. You still have some decisions to make. Don't worry, though; you have plenty of tools at your disposal. In the next sections, I guide you through some of the more common and powerful ones.

Determining the type of pattern

Figure 3-7 shows the distribution of tornadoes in Kansas from 1950 to 2023. Like the twister that dropped Dorothy's house on the Wicked Witch of the East, these tornadoes don't follow a regular pattern. But you don't need a wizard to figure out whether they're clustered, random, or something else entirely. You have GIS tools — no ruby slippers necessary!

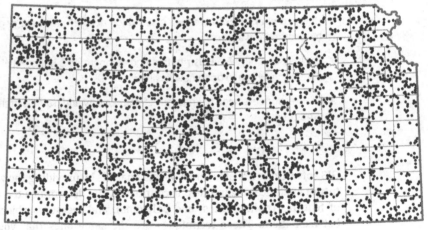

FIGURE 3-7: GIS can determine whether this tornado pattern is clustered or random.

Source: NOAA National Weather Service Storm Prediction Center, Severe Weather Database Files (1950-20223), https://www.spc.noaa.gov/wcm/#data

When points cluster, they tend to stick close to their neighbors, though a few may stray — like Dorothy and her friends strolling down the Yellow Brick Road. GIS uses the *nearest neighbor statistic* to help figure out if these points prefer hanging out together or keeping their distance.

To figure out whether a pattern is regular, random, or clustered, your GIS software follows these steps:

1. **It finds the random nearest neighbor distance.**

   Start by dividing one by twice the square root of the density of points. In the tornado example, the number of tornadoes in Kansas changes from season to season and year to year. So you need to measure the average nearest distance that would happen if the distribution were random. Use this simple equation:

   $$Random\ nearest\ neighbor\ distance = \frac{1}{2 \times \sqrt{density}}$$

   In this example, "density" is the number of tornadoes per square mile. This formula gives you the average nearest distance for a random pattern.

2. **The software then calculates the average pair-wise nearest neighbor distance.**

   The term *pair-wise* means comparing things two at a time. In this case, GIS matches, or *pairs*, each point (such as a tornado location) with its closest neighbor. It does so by finding the x- and y-coordinates for each point, measuring the distance to every other point, and determining which of those other points is the nearest. After collecting all the distances, the software calculates the average of these *pair-wise nearest neighbors*.

3. **It compares distances.**

   The software divides the average pair-wise nearest neighbor distance (from Step 2) by the random nearest neighbor distance (from Step 1). This comparison shows how your distribution compares to randomness.

4. **It interprets the number.**

   Compare the result from Step 3 to 0 and 1, where 1 means a perfectly random pattern and 0 means a perfectly clustered pattern. Numbers greater than 1 indicate a more regular, or dispersed, pattern.

If you calculate the average of all the nearest neighbor distances and divide that number by the random nearest neighbor distance — and you get a value of 1.0 — the pattern is perfectly random. If the value you get is 0.0, the pattern is perfectly clustered or as clumped together as you can get. Values greater than 1 show increased regularity, or they're more dispersed as the number gets larger. No absolute cutoff exists for "perfectly regular" distributions because the values you get depend on the point density.

TECHNICAL STUFF

You can go on to determine whether the values you get from your analysis occur by chance or whether they're truly meaningful, but you may want to leave that job for a statistician. You need to understand only the nature of the distribution.

TIP

The nearest neighbor analysis is an effective tool for points, but you can also use it to study the distribution of lines, areas, and volumes. Just let the software calculate the center of each line — or each polygon or volume feature — and then run the same calculation that you would for points. This analysis helps determine that what you see with your eyes is real and not a figment of your imagination.

Choosing an appropriate tool

GIS includes a variety of tools to help you identify and analyze different spatial patterns in your data. Here are a few of the more popular tools available in most GIS software applications:

» **Direction Analysis:** Curious about the average direction of tornado paths or how fallen trees align? This tool gives you a quick summary of which way these features typically point. It can reveal interesting insights, like the direction of prevailing winds or the path of glacial movements.

» **Network Analysis:** Need to explore the connectivity of roads, rivers, or other networks? Network analysis can help calculate things like ratios of stream branches or the circuitry of transportation networks to help you understand how these networks link to one another.

» **Density Analysis:** Ever wanted to spot hotspots? Density analysis creates heat maps to show concentrations, whether of population clusters, crime hotspots, or the spread of diseases.

» **Sinuosity Calculation:** Rivers rarely flow straight, and this tool measures just how twisty they are. Sinuosity gives you insights into the winding nature of rivers and streams, shedding light on how landscapes form.

Choosing the right tool depends on the patterns you're curious about and the data you're working with. Most GIS software comes with a variety of analysis tools, so don't hold back in exploring the help menu or provided tutorials; they often include a bunch of useful tips and examples.

Interpreting the Results and Making Decisions

Maps reveal all kinds of patterns and distributions. The more maps you read, the quicker you become at spotting distinctive patterns based on the layout, distribution, or location of features. GIS takes pattern spotting a step further by providing tools to measure and categorize these patterns, confirming what you see in the data. It's almost like being a detective: Tools in hand, you uncover the story behind the data.

Patterns form because of underlying forces, whether natural, human, random, or planned. When you start spotting these patterns, your next step is to figure out why they exist. Every pattern has a process behind it, and that process shapes the pattern. For example, in a forest, as trees grow and reseed, the density of trees increases. As this happens, competition for space leads to some trees dying off, changing the pattern of trees over time.

Here's a simple process to use for analyzing your GIS data:

1. **Spot the patterns.**

   Recognize that patterns exist in your data.

2. **Verify the patterns.**

   Use GIS tools to analyze and confirm these patterns.

3. **Understand the causes and effects.**

   Use your knowledge of the study area to figure out what's driving these patterns and what they mean.

4. **Apply your insights.**

   Use what you learn to make predictions and decisions.

GIS helps you with every step in this process except one: It can't tell you why the patterns occur. That's up to you. As the expert, you decide which factors to explore and which patterns to compare. GIS is an excellent tool, but like any tool, it's only as good as the person using it.

WARNING

GIS specialists may not always be the subject-matter experts. That's why collaboration is key to fully understanding patterns and their implications. For example, if a map shows a shift in traffic patterns, a traffic expert can interpret what that shift might mean (such as congestion possibly indicating the need for new roadways), whereas the GIS analyst handles the data analysis and visual display. It's a team effort!

2 Geography Goes Digital

IN THIS PART . . .

Find out how a conceptual model helps you lay the groundwork for your GIS project.

Understand the data models that GIS software uses to organize and manage spatial information.

Discover how to organize and manage your GIS data efficiently and tap into new database technologies.

Find out how to identify and obtain quality data for your GIS projects.

Chapter 4

Creating a Conceptual Model

IN THIS CHAPTER

» Compensating for your computer's lack of map-reading skills

» Understanding how to select the right data for your GIS project

» Figuring out what your GIS project needs

» Choosing an appropriate data format

Maps are packed with stories. They contain information hidden within their symbols, lines, colors, and shapes. For example, a map with different shades for soil types doesn't just show where different soils are located; it tells you which soils are common, which are rare, and how they're spread across your study area. Maps do more than present facts; they reveal relationships, connections, and patterns in the world around you.

Take, for instance, a map with point symbols representing fire-lookout towers. Each symbol links directly to a tower's absolute location, but the map tells you more. You can see the distance between towers, how well they cover fire-watch zones, and how they're distributed across the landscape. The story within the map tells you more than what's immediately visible.

Maps are complex, and humans like you and me have been taught how to read and interpret them. GIS is just software installed on a computer. A computer doesn't know how to read a map the way you and I do unless we teach it to recognize patterns and interpret information the way a human does. That's where a conceptual model comes in.

A *conceptual model* is like a mental blueprint for how you teach the computer to display and analyze map information. In this chapter, you discover how to create a conceptual model that helps you translate your map-reading instincts into GIS projects that reveal the stories hiding in your data.

Helping Computers Read Maps

Maps tell a story. Squiggly contour lines drawn tightly together convey a steep slope. In another spot, where they're more spread out, they describe a gentle incline. When you spot a fire lookout symbol on the map with tight contour lines around it and a winding trail leading up to it, you instinctively know that people hike uphill to reach it. It's second nature to you, right?

Unfortunately, your computer doesn't have a "second nature" instinct. It doesn't understand anything about your map. To your computer, those contour lines are just numbers in a file. It doesn't know that those lines represent elevation or that symbol means "fire-lookout tower." And it definitely doesn't understand the hill or the uphill hike! When it comes to reading maps, computers are . . . well, dumb.

That's where you come in. To help the computer, you need to build a *conceptual model*. It's kind of like teaching your computer to "read" maps. You break down the map's structure, decide what features are important, and explain how they're connected. With these details in place, you create a model that your computer can understand. Now the computer can recognize that a fire lookout tower sits on a hill with a trail going down the slope to the trailhead.

Embracing the Model-Creation Process

Before diving into any GIS project, knowing what you plan to create is important. Think of it like planning a road trip. Unless you're really adventurous, you don't jump in the car and drive off without knowing your destination. You start planning your road trip with the big picture, typically by knowing where you want to end up. It's the same with a GIS project. Start by asking yourself, "What is my end product?" With that vision in mind, you can start breaking it down into smaller pieces, *or themes*, like planning the stops on your road trip. Finally, you figure out the specific details for each piece by defining the map *elements* — the features you need to highlight, and the data that will bring all the elements together.

TIP

Each GIS project is unique. I suggest going analog first and using a pen and paper. Decide what features you need and sketch out a simple flowchart, starting with the end goal and working backward toward each individual feature. Don't worry about data availability; imagine you have it all at your fingertips. You can come back and modify your flowchart later, after you know what data you can actually access.

Defining Your Map's Contents

Before you create a map, whether it's *analog* (hard copy) or *digital* (electronic), you need a sense of the geography you want to include. What elements you put on the map and into your GIS can make or break your analysis.

Whether you're working with economic geography, medical geography, transportation geography, urban geography, or something else, each type has its own specific data and analysis needs. For example, suppose you're using GIS to improve the productivity of your stores. You might consider the following:

- **Economic geography:** What is the average income of shoppers in your area?
- **Transportation geography:** How do people get to and from your stores?
- **Physical geography:** How do seasonal weather changes impact what customers buy?

Your mission, should you choose to accept it, is to pick the geographies that best fit your project and use them to build the perfect maps.

Choosing the data

I use a simple trick to decide what data I include in my GIS project. I always start at the end, outlining what I expect the final product to look like. For example, if you're using GIS to find the best location for a new housing development, you might ask yourself these questions:

- How big of a piece of land do I need?
- How much am I willing to pay for the land?
- Are there any locational constraints? Where do I prefer the land to be?
- What are the legal constraints — such as zoning and ownership — for developing the land?

CHAPTER 4 **Creating a Conceptual Model** 59

» What physical constraints am I willing to accept?

» What are the infrastructure needs for this development?

Your answers to these questions will help you determine the data you'll need for your GIS project. The data will most likely include:

» Parcel data to determine those that fit within your size requirement.

» Current sale price or value of the land to find the parcels within your price range.

» Important features such as hospitals, schools, and parks so that you can measure proximity to them.

» Local zoning data to determine whether the land is zoned for residential development and you'll need to get a zoning variance to build there.

» Land ownership data to know from whom you need to purchase the land, or even whether you can buy it at all.

» Topography, soils, vegetation, bedrock, and other physical factors to decide how much work you have to do to develop it.

» Infrastructure data such as power, water, and broadband to determine whether you need to invest more to have these installed.

TIP

By starting at the end, you break your project into manageable pieces. For example, asking the right questions about what you need for a new housing development lets you translate those questions into specific map features, like parcel size, zoning, and infrastructure availability. This step-by-step approach keeps your project on track and ensures that you gather the right data for your analysis.

Applying the methodology

Even if your project isn't related to housing development, the principles outlined in the previous section, "Choosing the Data," still apply. Whether you're selecting the best location for a new store, determining the best area for a wildlife preserve, assessing crime hotspots, or predicting the risk of a hazardous cargo spill, the question is the same: *Where?* Location is the core of every GIS project.

Every major GIS project revolves around the locations and distributions of features. Sometimes you use GIS to describe the impact of existing locations; at other times you make decisions about where to place businesses, public services, or research facilities. Whatever your application, the methodology is the same: Define what you need and break down the data into manageable pieces.

60   PART 2 **Geography Goes Digital**

No matter the topic, the best place to start is the end. When you visualize the end result, you identify the relevant parts of the project and plan accordingly.

Breaking down the data

Each data layer that you select for use in your GIS database includes not just the geographic features but also the information, or *attributes*, tied to those features. Sometimes you need all the attributes associated with a particular feature; at other times just a few will do.

Consider the housing development example in the "Choosing the data" section, earlier in this chapter. A soils data layer might contain useful attributes for determining whether the soil can support construction. But it may also include details like the soil name or classifications that aren't relevant to your project.

Next, outline the details of your project and ask specific data-related questions. Your housing development project involves more than just the land itself. Think about all the requirements for your ideal parcel, such as size, price, location, and access to utility infrastructure. Breaking these details into smaller chunks makes them easier to work through.

Suppose you're also looking to create a bird sanctuary near the development. You need land that meets the unique requirements for that. Here are some data layers that can help:

» **Land ownership:** Use this layer to identify who owns the land. You most likely need to exclude government-owned parcels and focus on private land that's available for purchase.

» **Land for sale:** This layer shows you which parcels are for sale, their sizes, and (you hope) their prices. Your goal here may be to find a single 20-acre parcel for under $2,000 per acre.

» **City limits:** You want people to visit the sanctuary, so it will need to be within five miles of the town. A boundary data layer of the city limits shows you how far each parcel is from the city.

» **Transportation:** Being close to town doesn't help you if people have no way to get there. Use this layer to ensure access via two-lane roads or highways.

» **Utilities:** You need electricity, gas, and water at the sanctuary. This layer allows you to see how far away each parcel is from these utilities.

CHAPTER 4 **Creating a Conceptual Model** 61

Figure 4-1 shows a flowchart of the process you use to break down your project details. The final product (the bird sanctuary) is on the right, and its components are on the left. Table 4-1 breaks down these project details and identifies the questions you ask to pinpoint the data you need.

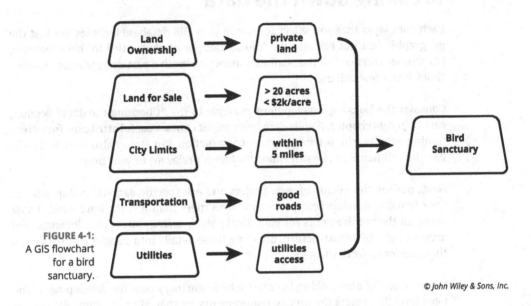

FIGURE 4-1: A GIS flowchart for a bird sanctuary.

© John Wiley & Sons, Inc.

TABLE 4-1 Selecting Data from Spatial Data Layers

Project Detail	Data Layer	Questions to Ask	Data to Select
Land that can be purchased	Land ownership	Is the land publicly or privately owned?	Privately owned land
Land that's suitably sized and priced	Land for sale	Which parcels are for sale? How big are the parcels? How much do they cost?	Parcels for sale Size is at least 20 acres Cost is less than $2,000 per acre
Land that's appropriately located	City limits	How far are the parcels from the city limits?	Parcels within five miles of city limits
Land with access to adequate transportation	Transportation	What kinds of roads provide access to the parcels?	Parcels accessible by two-lane or wider roads
Land with public utilities available	Utilities	What utilities are available for which parcels?	Parcels that have access to public water, electricity, and gas

62 PART 2 Geography Goes Digital

Verifying data characteristics

Each data layer in your GIS project consists of features like points, lines, areas (polygons), or volumes, all of which contain valuable information. Early in your GIS process, it's important to understand the data you have and their key characteristics:

» **Level of measurement:** You can measure data on different scales: nominal, ordinal, interval, or ratio. The level of measurement influences how you interpret and use the data. For example, nominal data represent categories, whereas ratio data involve measurements with a true zero point. (See Chapter 3 for an in-depth description of these measurement scales.)

» **The level of detail:** The scale at which you map data determines how much detail is available. Larger-scale maps (covering smaller areas) provide more detail, whereas smaller-scale maps (covering larger areas) simplify features. For instance, a feature that appears as a point on a small-scale map may show up as an area on a large-scale map. (Check out Chapter 2 for more on the importance of scale.)

Starting your GIS project with the most complete and detailed data possible gives you more flexibility in your analysis. My basic rule of thumb is to opt for detailed data whenever possible because generalizing from detailed data is easier than trying to extract specifics from generalized data.

Converting Maps into Digital Formats

GIS helps you analyze your study area by mapping it digitally, but before you jump in, you need to plan your data. To answer your project questions, you include specific features like points, lines, areas, and surfaces in your map. But here's the thing: Your computer isn't as smart as you are when it comes to understanding the data.

Deciding how to represent your data

GIS uses two main methods for storing and displaying map, or *spatial*, data: *raster* (pixels or grid cells) and *vector* (points, lines, and polygons). (For more on the basics of raster and vector data, see Chapter 2.) Both data types have their pros and cons, but most GIS software can handle both, so you don't need to panic about which to use.

CHAPTER 4 **Creating a Conceptual Model** 63

Picking a raster-based model

Think of a raster data model like a grid of tiny squares, where each represents a piece of the Earth. Imagine a checkerboard with each square, or *cell*, covering part of the landscape. Use this grid model to represent different types of features:

>> **Points:** In the real world, points have no length or width (zero dimensions). In raster data, points are represented by a single grid cell. For example, a fire-lookout tower might appear as one colored square in the grid. Although points are simplified in raster format, they can still represent critical locations like water wells or city landmarks. The top portion of Figure 4-2 shows how points might appear in a raster.

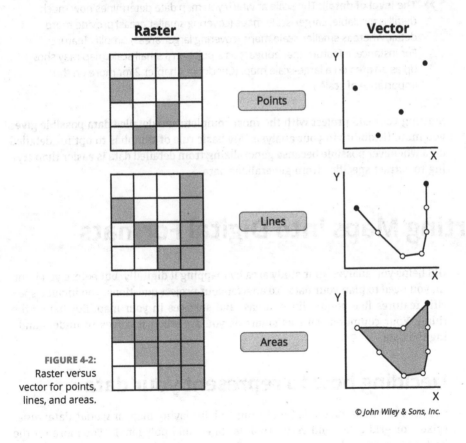

FIGURE 4-2: Raster versus vector for points, lines, and areas.

© John Wiley & Sons, Inc.

PART 2 **Geography Goes Digital**

- **Lines:** Although lines in real life may have width (such as wide highways or rivers), raster data simplifies them into strings of grid cells that show their central path. As Figure 4-2 shows in the raster examples on the left, the cells can connect *orthogonally* (side to side, top to bottom, or edge to edge) or *diagonally* (corner to corner) depending on the shape of the line. If necessary, you can capture the width of larger features by grouping adjacent cells. For example, wider rivers or highways might be represented by groups of cells, whereas narrower ones may appear as a single row of cells.

- **Areas (polygons):** Areas, also called polygons, represent larger spaces like fields or lakes. In raster format, areas are made up of multiple adjoining grid cells, and each cell is assigned a value representing the type of land use, like forest, agriculture, or water. For example, crop areas may be displayed as blocks of cells. Figure 4-2 shows how a particular crop area can look in a raster.

- **Surfaces (volumes):** Surfaces represent the Earth's 3D features, such as elevation or groundwater depth, using a group of cells. Each cell covers a specific area on the ground and contains a value representing height (elevation) or depth. A classic example of this is a Digital Elevation Model (DEM), a widely used rater dataset in which each cell stores an elevation value, allowing you to analyze terrain features. Think of this representation like a blanket draped over the terrain, with each square on the blanket corresponding to a section of land. The colors or shades applied to the cells indicate ranges of values (such as elevation). For instance, higher elevations might appear lighter, whereas lower areas are darker, as shown in Figure 4-3. This grouping of shaded cells helps you visualize gradual changes in the terrain, such as valleys and peaks. Even though this grouping simplifies the view, GIS keeps track of the exact value for each cell, preserving accuracy. Storing the values within each cell allows you to zoom in and measure surface features at any location on the map.

REMEMBER

Raster data represent a point with a single grid cell, a line with a line of grid cells, an area with a group of grid cells, and a surface with groups of grid cells that have additional unique values.

Choosing a vector-based model

Vector models work differently than raster models. Instead of dividing the world into tiny grid cells like a raster model, vector data represent features like points, lines, and areas (polygons) using precise X and Y coordinates, thereby making them ideal for capturing exact locations and boundaries. Here's how each feature type is handled in the vector model:

» **Points:** A point in vector data is mapped by a single X and Y coordinate, like placing a pin on a map. A point has no measurable length, width, or area. For example, you can represent features like trees, landmarks, or fire-lookout towers this way, with each having a specific location without taking up space on a map.

» **Lines:** Lines represent linear features, such as roads, rivers, or pipelines. You form lines by connecting two or more points (X, Y coordinates). In contrast to raster models, vector lines can capture intricate curves or straight paths with high precision. If the width is important, such as with roads or rivers, you can use additional attributes to store this detail or use a polygon to represent the feature. Refer to Figure 4-2 to see how lines are represented as a vector.

» **Areas (Polygons):** Areas, also known as polygons, are shapes formed by closed lines. For instance, a lake, a city boundary, or a parcel of land is represented as a polygon. You create these shapes by connecting three or more points, with the start and end points meeting to form a closed boundary. Polygons are great for representing things that occupy space, like parks or building footprints.

» **Surfaces (Volumes):** Representing surfaces in vector form is a bit more difficult than with points, lines, or polygons. To handle this task, GIS uses a method called *Triangulated Irregular Network (TIN)*, which breaks a surface into triangles, with each triangle's corners (vertices) assigned a height (Z value) in addition to the X and Y coordinates. TIN excels at detailed terrain modeling, whether you're mapping elevation in a mountainous region or calculating slope for a potential construction site. Refer to Figure 4-3 to see how TIN models compare to raster.

Weighing the benefits: Raster versus vector

Raster or vector? You may wonder how to decide which type of data to use in your GIS project. GIS software handles both data models well, so the answer really comes down to the specific needs of your project. Here are some considerations to take into account:

» **Space:** Raster data takes up more storage because each grid cell must store a value. The size of these cells affects both storage and detail: Smaller cells provide higher resolution but require significantly more storage; larger cells reduce file size at the cost of spatial detail. Vector data, on the other hand, can represent large areas efficiently with just a few X and Y coordinate pairs.

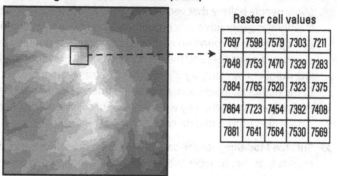
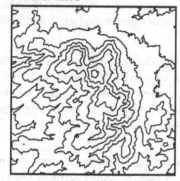
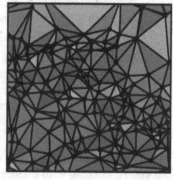

FIGURE 4-3: Surfaces can be represented as raster, vector, or TIN.

© John Wiley & Sons, Inc.

» **Speed:** When you're dealing with large datasets, processing raster data is typically faster than vector, especially when performing processing tasks like map algebra (explained in Chapter 16). Vector data, on the other hand, may take longer when processing large datasets or complex geometries but is better suited for tasks requiring high precision.

» **Compatibility:** Raster is better suited for working with satellite imagery or scanned maps because it's already in a similar grid format. Vector is more compatible with detailed line drawings or computer-aided design (CAD) files.

» **Accuracy:** Vector data are generally better for representing distinct boundaries, such as property lines or roads, because it records exact coordinate points. Raster data, however, is better for representing continuous phenomena like elevation or temperature, whose values change gradually across a surface. Although smaller raster cells capture more variation and greater accuracy, they still generalize features compared to vector representations.

WARNING

Many people believe that vector data gives an accurate representation of locations in geographic space, but that isn't entirely true. Computers sometimes truncate or round up numbers, have single or double *precision* (the number of points beyond the decimal point that they round to), and manipulate numbers by using algorithms that do strange things (other than rounding) to the numbers. In addition, all data, whether vector or raster, are only as accurate as the method by which they were collected. See Chapter 2 for details on methods of data collection.

» **Surface Modeling:** Raster data is perfect for modeling things that change gradually across surfaces like temperature, pollution dispersion, or rainfall. Each raster grid cell stores a value, so the entire surface is covered without any gaps. Complete coverage with a value in each cell makes raster data the best choice for environmental modeling and working with large-scale imagery. Vector data (such as a TIN) can also model surfaces, but because it uses irregularly shaped areas instead of a grid, it can be harder to work with for very large or complex datasets.

» **Visual Appeal:** Vector data stays clear and accurate regardless of zoom level, making it better for maps that need to be displayed at various scales, such as city maps or utility networks. In contrast, raster data can look blocky, especially with coarser grid sizes. If you zoom too far into a raster, you might notice that features start looking pixelated.

Ultimately, choosing between raster and vector comes down to your project needs. If you're working with imagery or need to model surfaces like elevation, raster is the way to go. For detailed, precise maps of roads, parcels, or city boundaries, vector data will serve you better. Most GIS projects involve a mix a both!

> **IN THIS CHAPTER**
> » Finding your way around a raster representation
> » Exploring vector models
> » Understanding raster and vector surfaces

Chapter **5**

Understanding the GIS Data Models

When you read a map, you're seeing a visual representation of the real world. As a GIS analyst, though, your job goes beyond simply reading maps. You need to understand how to combine, analyze, and make decisions based on the map's representation of the real world. The way GIS stores and represents data determines what you can do with it.

Unlike a paper map, GIS doesn't just store an image representing a location on the Earth. Instead, it structures data in ways that allow you to perform analyses, queries, and modeling. To use GIS effectively, you need to understand the underlying data models — the frameworks that tell the software how to organize and manage spatial information.

In this chapter, you explore the core GIS data models, their strengths and limitations, and how they make sense of the real world. Knowing how data models work can help you unlock the full potential of GIS for your projects by making the jump from simply viewing the map to truly understanding and using its data.

Examining Raster Models and Structure

Raster data models consist of a grid of cells, like a checkerboard. Instead of black and red squares with a round checker piece in it, however, each square in a data model represents a tiny unit of geographic space. On a raster "game board," GIS assigns every square, or *cell*, a value. These cells represent different geographic features such as roads, rivers, houses, or parks.

With checkers on a checkerboard, you can move the pieces around, but only in specific ways based on the game's rules. Similarly, raster grids follow rules about what can be where. Each square, or cell, in the grid has limits on what it represents, whether it's land, water, roads, or other features. Geographic space in a raster is divided into small, manageable cells, but each one must follow the "rules" of what it can depict.

The real world isn't divided into neat little squares as a checkerboard is. Geographic features like rivers, forests, and roads flow and overlap in irregular ways. A raster grid simplifies the complexity of dividing the Earth's surface into uniform cells, with each representing a specific value for that part of the landscape. However, this simplification comes with a trade-off: Although it makes analyzing geographic data easier, it also reduces detail. Small features may get "averaged out," and the shapes of natural features can look blocky or less precise. Figure 5-1 shows how this simplification works. The grid overlays the landscape, providing a way to analyze geographic patterns but with some loss of finer details.

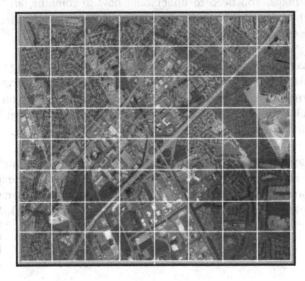

FIGURE 5-1: A grid overlay shows how each square represents a portion of real geographic space.

Representing dimension when everything is square

As mentioned in the previous section, raster models work like a checkerboard grid in that they consist entirely of squares. Every square, or *grid cell*, contains a piece of Earth's surface data. But the real world isn't made up of squares; it's a more complex arrangement of points, lines, areas, and even 3D surfaces (also called *volumes*).

When it comes to surfaces, things get a bit tricky because surfaces introduce a third dimension, like height or depth. You can dive into the third dimension in the section "Dealing with Surfaces," later in this chapter.

For now, the following list describes how points, lines, and areas fit into a raster grid system (as shown in Figure 5-2):

» **Points:** In the real world, a point represents a specific location, like a tree or well. A point is considered zero-dimensional because it doesn't take up space. In a raster model, even this tiny location takes up one full grid cell.

» **Lines:** A line is a one-dimensional object that represents a feature like a road or river. In a raster model, a line appears as a string of connected grid cells tracing the line's path.

» **Areas (Polygons):** An area, such as a park or lake, is a two-dimensional object. An area covers more ground and is represented by a group of connected grid cells that form its shape.

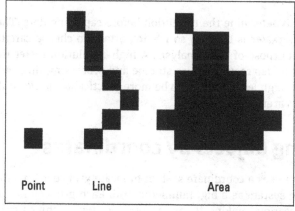

FIGURE 5-2: Point, line, and area features in the raster data model.

© John Wiley & Sons, Inc.

CHAPTER 5 **Understanding the GIS Data Models** 71

Each type of feature, whether a point, line, or area, fits into the grid in its own way. So although raster data divides everything into simple squares, it still captures these different geographic features effectively.

REMEMBER

Raster models make points, lines, and areas fit into grid cells: Points take up a single cell, lines are strings of cells, and areas are groups of cells. See Chapter 4 for more details on how grids represent geographic features.

Making a quality difference with resolution

At first glance, using two-dimensional squares (raster grid cells) to represent points or lines may not seem to make sense. How does one little tree fit into a big square? Well, it doesn't really, but raster cells simplify the data, helping your GIS store information about that one little tree in a structured and efficient way.

This simplification comes with trade-offs. Raster models divide space into a grid of uniform cells, and every feature within a cell — whether that feature is a single tree or an entire building — is represented by a single value. This means that some spatial detail and precision are lost, especially for points and lines. But in exchange, raster models provide an efficient and powerful way to store, analyze, and model geographic data.

When you work with raster data, one of the most important decisions you make is the choice of *grid cell resolution*, which is the size of each cell in real-world units (such as meters or feet). Smaller grid cells provide greater detail, a lot like how a high-definition TV screen has more pixels to create a sharper image. A raster with 1-meter resolution, for example, captures finer details than a raster with 30-meter resolution because each 1-meter cell represents a smaller area on the ground.

REMEMBER

You need to determine the resolution before capturing data. The grid cell size is set when a raster is created, so it's important to choose carefully based on the scale and purpose of your analysis. A high-resolution raster may provide more detail, but it also requires more storage and processing time. A lower-resolution raster, although less precise, can be more practical for large-scale analysis where fine details aren't necessary.

Finding objects by coordinates

Raster data uses a coordinate system to locate grid cells on a map. Think of this coordinate system as a big, numbered grid. Each grid cell has a column (X) and row (Y) location, making it easy to pinpoint where things are. The numbering typically starts at the upper-left corner of the grid and goes from left to right and top to bottom. But the numbering can just as easily start from the bottom left and move up. As long as the grid is consistent, it works! Take a look at Figure 5-3 to see the structure of a typical grid system.

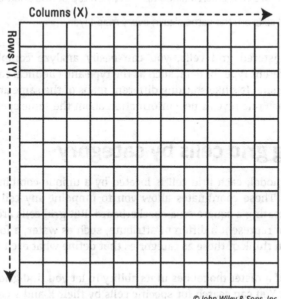

FIGURE 5-3:
A typical raster grid coordinate system.

Moving from one grid cell to the next represents real-world distance and direction. For example, if your grid cells are 100 meters on each side, each cell covers 10,000 square meters (100 x 100). If you move five cells to the right, you travel 500 meters east — simple as that! Similarly, moving down a row means traveling 100 meters south. (All these measurements assume that your grid is aligned in cardinal directions.)

With this system, you can find any object by its column and row coordinates. Even better, because these coordinates are tied to real-world latitude and longitude, you can link your grid cells to specific locations on Earth. This capability is when GIS gets fun!

When you layer grids for modeling, you can compare and analyze different features that share the same location on different map layers. For example, you might compare the location of a building (from a land-use layer) with its soil type (from a soils layer) and any hazardous chemicals in the soil (from a soil-chemistry layer). This analysis is possible because each layer uses the same grid system, and the grid cells line up perfectly across the layers.

For this process to work, though, you need to make sure that

- » All the grids cover the same area.
- » The grids are *co-registered* (lined up exactly on top of one another).
- » The grid cells are the same size in every map layer.

With co-registered grid cells, you can easily analyze complex relationships between different types of data, such as the type and chemical composition of soil at a building site. It's like stacking different maps of the same area on top of each other, but each one reveals new information about the same spot on the Earth.

Finding grid cells by category

In a raster model, each grid cell is located by a unique combination of X and Y coordinates. These coordinates allow you to pinpoint any cell within the grid, much like locating a square on a checkerboard. But grid cells aren't all the same. Each cell can represent a different attribute, such as water, a forest, or an urban area. You can think of these as categories that define what each cell represents.

The power of a raster model lies in its ability to let you find grid cells in two different ways. You can search for specific cells by their X and Y coordinates, which is useful when you need to locate a specific point on your map. Alternatively, you can search by category — say, all cells that represent sandy soil or all cells with a particular value — to easily identify and analyze areas that share similar properties.

REMEMBER

Raster GIS gives you two ways to search its grid:

- » **By coordinates:** Locate a grid cell based on its position in the grid.
- » **By category:** Find cells based on their assigned feature or attribute.

Working with map layers

Early GIS software used methods like the Map Analysis Package (MAP) to store and manage spatial raster data. Today, GIS software applications like ArcGIS, QGIS, and others handle these tasks through more sophisticated data-management methods, but the concept of organizing data by map layers remains foundational.

Modern raster systems represent data as layers, each with its own name and theme (for example, land cover, elevation, or soil type). You can think of each map layer as a separate piece of information about the landscape. You can then overlay or combine the layers to perform spatial analysis. If you need more details on

74   PART 2 **Geography Goes Digital**

these methods, I describe the raster overlay method in Chapter 15 and methods for combining raster layers in Chapter 16.

Here's how the components of a raster layer work:

- **Layer (Theme):** Each map layer contains a specific type of data, such as land cover, elevation, or soil type.
- **Category:** You can divide each layer into categories representing specific attributes like water, a forest, or an urban area in a land-cover layer.
- **Value:** GIS often stores categories as numbers to help with analysis. For example, water might be represented with a value of 1, a forest with 2, and an urban area with 3.

Figure 5-4 shows how land-cover categories might be represented in a raster layer.

1	1	1	1	2	2	2	2	2	2
1	1	1	1	2	2	2	2	2	2
1	1	1	1	1	1	1	2	2	2
1	1	1	1	1	1	1	2	2	2
2	2	1	1	1	1	2	2	2	2
2	2	1	1	1	2	2	2	2	2
2	2	2	2	2	2	2	2	2	2
2	2	2	2	2	3	3	3	3	3
3	3	3	3	3	3	3	3	3	3
3	3	3	3	3	3	3	3	3	3
3	3	3	3	3	3	3	3	3	3

Land Cover
1 = water
2 = forest
3 = urban area

FIGURE 5-4: Categorical data represented in a raster layer.

© John Wiley & Sons, Inc.

Linking objects with attributes

Modern GIS software enhances the capabilities of raster data by linking spatial features to a broader database of attributes, often using a relational database management system (RDMS). With an attribute table, a layer that represents land use can include more than just a housing category. Your land-use raster layer can also contain detailed information about different housing types, such as single-family and multifamily.

Software applications like ArcGIS Pro and QGIS provide tools for managing these linked databases, allowing users to access multiple values for each spatial category. This added flexibility means that you can conduct deeper levels of analysis. For instance, you can assess the environmental impact of housing types or evaluate patterns in housing density.

Extending the raster data model by linking it with a database management system gives you more flexibility. It allows for the representation of multiple attributes for category, making your analysis more dynamic. Be sure to use meaningful and memorable names for categories, ID codes, and map layers to keep everything organized and easy to retrieve.

Exploring Vector Representation

Think of vector data as digital building blocks of the world, like the points, lines, and polygons that you would draw manually on a map. Each point has X and Y coordinates, with lines connecting these points in sequence and polygons looping around and ending where they started.

But vector data doesn't stop with simple points, lines, and polygons. How does a road "know" it connects to another road? Or how does one town boundary "know" it neighbors another? This section explores how GIS turns these simple shapes into intelligent, interactive maps that understand how the world fits together.

Moving beyond a simple model

The *spaghetti model* is one of the earliest models used in GIS for handling vector data. In the early days of GIS, this model provided a quick and easy way to store points, lines, and polygons independently. It earned its name because the data structure resembles a tangled pile of noodles on a plate — (although not smothered with red sauce and meatballs). In the spaghetti model, each feature stands alone, with no awareness of other features around it, as shown on the left in Figure 5-5. Although the model was great for input and output, it quickly became impractical for today's complex datasets in which relationships between features matter.

Today, spaghetti models have largely fallen out of use. Instead, modern GIS relies on topological models (explained in the next section) because they store spatial relationships. In these models, your GIS software understands which polygons share borders, which streets connect at intersections, and how areas nest within one another, thereby making analysis far more powerful and accurate.

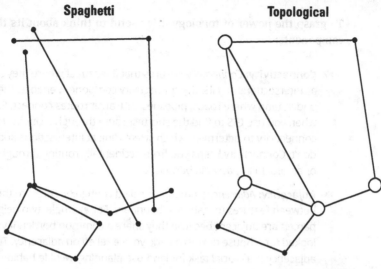

FIGURE 5-5:
A spaghetti model compared to a topological model.

© John Wiley & Sons, Inc.

Adding spatial smarts to GIS with topology

Topology is a branch of mathematics that describes how features (geometric objects) connect — or don't. GIS uses topology to define and enforce spatial relationships, ensuring that features align and connect correctly in the database. Topology is the secret behind GIS's ability to recognize shared boundaries between polygons or connections at street intersections. Just as your mind uses spatial reasoning (your sense of space) to understand that the coffee shop is next to the grocery store, topology gives GIS a set of rules that explicitly define these relationships — making spatial analysis more accurate and producing more reliable results.

In the real world, features interact, meaning that rivers flow along county boundaries and roads meet at intersections. GIS needs topology to understand these relationships and prevent errors such as the following:

» Roads don't connect where they should.

» Gaps exist between property boundaries.

» Polygons overlap when they shouldn't.

REMEMBER

When topology is included in the data, GIS can quickly validate and analyze spatial relationships. Validation and analysis are especially important for tasks like network analysis (finding the shortest route to the coffee shop) or property management (making sure that parcels align correctly).

CHAPTER 5 **Understanding the GIS Data Models** 77

To grasp the power of topology, it's useful to think about its three fundamental components:

- **Connectivity:** *Connectivity* ensures that lines meet where they should, such as intersections. In GIS, the connectivity component enables the software to understand where roads, pipelines, or transit routes connect. For example, when you use GIS to find the shortest route through a city, you rely on connectivity to determine which streets link at intersections and which ones don't. Connectivity keeps you from accidentally routing through a dead end or across an impassable barrier.

- **Adjacency:** *Adjacency* (sometimes called contiguity) refers to the relationship between features that share a boundary. For example, two neighboring land parcels are adjacent because they share a common border. If you've ever looked for a house next to a park, you've relied on adjacency. The concept of adjacency is a crucial task for land-use planning, wildlife habitat connectivity, and any analysis that depends on knowing what's next to what.

- **Containment:** *Containment (also known as area definition)* ensures that boundaries connect to create closed shapes, forming polygons. With well-defined polygons, GIS can calculate metrics like area size or perimeter. This capability is important for assessing how much space a city park covers within the city's boundaries or determining the total area of a lake inside a watershed boundary.

Topology isn't just a fancy term for how features relate to each other; it plays a key role in keeping GIS data accurate. Without topology, roads may not meet at intersections, property boundaries can overlap, or pipelines may fail to connect (at least in your GIS; if they aren't connecting in the real world, you've got a bigger problem!). Imagine trying to model a city's water system if the pipes didn't link properly. Topology ensures that everything connects as it should. Refer to Figure 5-5 to see how the topological model compares to the simplistic spaghetti model.

TECHNICAL STUFF

Many modern GIS systems enforce topological rules through advanced data models, like geodatabases and PostGIS-enabled spatial databases. These systems help users automatically validate and clean data, making analysis smoother and more reliable. For example, geodatabases ensure that features like water pipes connect only at valid points (called *nodes*), which reduces errors in utility networks. PostGIS adds spatial capabilities to PostgreSQL, enabling powerful queries based on spatial relationships like "adjacent to" or "within."

REMEMBER

Although not every project needs topology — simple maps or visualizations can skip it — topology is essential when data accuracy matters, such as in infrastructure management or for property records. A strong topological structure allows GIS to help you find answers to questions like "Which parks are within this county?" or "Which rivers flow into this lake?"

Using shapefiles for easy data exchange

The shapefile is the GIS equivalent of the "lowest common denominator" format, being easy to use and widely supported by most spatial software. A *shapefile* stores geometry — like points, lines, and polygons — along with descriptive attributes, making it a practical format for many GIS users. Shapefiles are lightweight, fast to render, and compatible with most GIS platforms.

But don't be fooled by the name; a shapefile is actually a bundle of files. For starters, a shapefile has the three files at its core, listed in Table 5-1.

TABLE 5-1   **The Core Files of a Shapefile**

File Type	File Extension	What It Does
Shape format	.shp	Holds the geometry (the points, lines, and polygons)
Shape index	.shx	Contains an index of the geometry to speed up access
Attribute table	.dbf	Stores the tabular data associated with shapes (the descriptive data)

Table 5-2 describes some of the companion files that a shapefile usually includes.

TABLE 5-2   **Common Files in a Shapefile**

File Type	File Extension	What It Does
Projection	.prj	Defines the coordinate system and map projection (like Universal Transverse Mercator or UTM). Without this file, your GIS software may not properly align the shapefile with other datasets.
Code page	.cpg	Specifies the text encoding used in the attribute table (like UTF-8 or ANSI). This file ensures that non-English characters display correctly.
Metadata	.xml	Contains information about the shapefile, such as its creation date, data sources, and attribute descriptions. Some software, like ArcGIS, automatically generate this file.
Spatial index	.qix	An index that speeds up some spatial queries. Some software, like QGIS, generate this file automatically when working with large datasets
Attribute index	.atx	Stores an index for a specific attribute field, speeding up attribute-based queries (used by ArcGIS).
Spatial index for features	.sbn/.sbx	These files store a spatial index, a data structure that organizes spatial queries to improve performance, especially useful when working with large datasets.

CHAPTER 5   Understanding the GIS Data Models   **79**

If you're sharing a shapefile with someone, the fastest way to their heart is to include all the associated files, not just the .shp. A complete shapefile isn't just one file; it's a bundle. So if your dataset is called LandUse.shp, you'll also need to send all the related files like LandUse.shx, LandUse.dbf, and landUse.prj (and any others in the same folder). Skipping any of these files can result in confused or emails from your annoyed GIS friend, and nobody wants that!

Shapefiles are great for basic tasks but they come with serious limitations:

» **No topology:** They don't store relationships between features, such as which polygons share borders. I explain more about topology in the next section.

» **Limited attribute storage:** Field names can be only ten characters long, which can be restrictive when working with detailed datasets. If you're converting from another format, longer names get truncated, which can make it harder to identify fields later. Additionally, shapefiles don't support null values for numeric fields.

» **2GB size limit:** Shapefiles can't handle large datasets, which is frustrating for national-level or detailed datasets.

» **Precision issues:** Shapefiles store coordinates with limited precision, leading to rounding errors in high-resolution datasets.

When it comes down to it, though, shapefiles are like an old pair of shoes: You can rely on them when needed because they get the job done, quirks and all.

Fortunately, you have better options than shapefiles. One is a GeoPackage, which offers the same portability but with more flexibility. It stores geometry and attributes in one file, supports topology, handles larger datasets, and works with both desktop and web-based GIS tools.

If you need even more capabilities — such as for managing multiple datasets, defining relationships between features, or performing network analysis — consider using a geodatabase. *Geodatabases* (like those supported by ArcGIS) allow you to store raster, vector, and tabular data together, along with rules that govern how data interacts (like preventing invalid road connections). They're ideal for large-scale projects for which you need both data integrity and efficient searches.

Shapefiles are still useful for simple data exchange and lightweight mapping, but for more complex projects, a GeoPackage or a geodatabase will serve you better. As GIS continues to evolve, these newer formats are replacing shapefiles — but it's always good to keep an old pair of shoes around, just in case.

Building smarter systems with object-oriented models

Modern GIS systems rely on object-oriented programming (OOP) to make data smarter and more manageable. In these systems, each geographic feature is treated as an object that combines *attributes* (like name or length) and *behaviors* (rules about how the feature interacts with other features). For example, a road object knows it's a road, follows network rules, and won't connect to an incompatible feature, like a river (see the earlier section, "Adding spatial smarts to GIS with topology," for more on enforcing spatial relationships).

The geodatabase takes this approach further by storing diverse data types, like vector and raster, in one place, along with rules that govern their behavior. For instance, a geodatabase might ensure that one-way streets allow travel only in one direction and connect properly to other streets, preventing errors in network analysis.

TIP

Esri, the largest and most recognized GIS software company, continues to provide industry-specific geodatabase models, particularly for areas like utilities, transportation, public safety, and natural resources. These prebuilt models are still helpful for organizations that want a structured starting point without building everything from scratch. However, Esri has been expanding its focus to cloud solutions (like ArcGIS Online and Enterprise) and interoperability with other systems, so users may use these models in combination with cloud-hosted databases.

TIP

If you're looking for free, flexible, and scalable solutions, the open-source GIS community is thriving. QGIS works very well with PostgreSQL/PostGIS to build custom geodatabases. PostGIS continues to be a powerful option for projects that require advanced spatial queries and network analysis.

Dealing with Surfaces

GIS data models do more than store points, lines, and polygons; they also represent elevation and terrain. When you need to analyze the shape of the land, GIS represents surfaces using either raster grids or vector-based structures. If you checked out the earlier sections in this chapter on raster and vector models, you already know that rasters store data in a grid, whereas vectors use points and lines to define features. The same principles apply to surface models: Raster-based surface models, like Digital Elevation Models (DEMs), store elevation values in a grid, whereas vector-based surface models, like Triangulated Irregular networks (TINs), use connected triangles to represent changes in terrain. These surface models help with everything from mapping hills to analyzing slopes and water flow.

Imagine that you're hiking up a mountain. Getting to the top is not just about following a trail marked on a flat map; you also need to know the ups and downs of the terrain. That's where surface modeling in GIS comes in. It adds that third dimension, giving you a full picture of the landscape, bumps and all!

Storing surface data in a raster model

The raster data model is one of the easiest ways to store surface data. Each grid cell represents not only an area on the Earth's surface but also a single elevation value. So a cell located at column X and row Y stores an elevation value (Z) for that specific area. (To see a visual representation of how rasters store elevation data, refer to Figure 4-3 in Chapter 4.)

Although the raster model is simple to use, it has a few limitations:

» **Limited accuracy:** Each grid cell stores only a single value for the whole area it covers, which can reduce precision.

» **Coarser detail with larger cells:** As grid cells get larger, you lose detail, making it harder to capture fine features in the terrain.

On the plus side, raster surfaces require less computational power, making them efficient and great for large-scale surface modeling. They're also easy to convert into other formats, like contour lines or topographic data.

Ever wonder how those beautifully detailed 3D topographic maps are made? The answer often starts with a Digital Elevation Model (DEM). DEMs store elevation data for regularly spaced points across the terrain, typically in a raster format. Each DEM cell has an X and Y coordinate representing a location on the Earth's surface, and a Z value depicting elevation at these locations. The U.S. Geological Survey (USGS) was one of the first major organizations to develop and distribute DEMs, and its datasets remain widely used today. You can use DEMs to create 3D models, slope maps, or even hydrological models.

TECHNICAL STUFF

DEMs come in several sizes, depending on the maps they were derived from. USGS DEMs are available in

» **7.5-minute DEMs (covering 7.5 minutes of latitude by 7.5 minutes of longitude):** These provide the highest level of detail and are commonly used for detailed terrain analysis.

» **15-minute, 30-minute, and 1-degree DEMs:** As the area covered increases, both horizontal and vertical accuracy decrease.

Smaller intervals like the 7.5-minute DEM are best for detailed studies, whereas coarser intervals like the 1-degree DEM are better suited for regional or broad-scale analysis.

You can download DEMs directly from the USGS website or purchase enhanced versions from commercial vendors. Most modern GIS software supports DEMs and can convert them into different formats, such as

» Raster grids for terrain modeling

» Contour lines to show elevation changes

» Vector topographic data for detailed mapping

This flexibility makes DEMs an essential tool for both terrain analysis and visualization.

Representing surfaces in a vector model

Vector-based surface models are more complex but offer higher precision. The most common vector surface model is the Triangulated Irregular Network (TIN). Think of a TIN as a surface made up of a bunch of connected triangles. These triangles represent flat parts of the terrain, and by connecting them edge to edge, the model creates a continuous surface. See Figure 5-6 to get a clear picture of how TINs connect triangles to represent elevation.

The TIN model breaks surfaces down into three components:

» **Triangulated:** You form each triangle using three points with known X, Y, and Z values.

» **Irregular:** The triangles vary in shape and size to match the terrain's complexity.

» **Network:** The triangles connect along their edges to form a seamless surface (known as a facet).

Each triangle's slope and direction (aspect) remains consistent within the triangle, which makes TINs ideal for capturing small, detailed variations in terrain. In the real world, engineers often use TINs when planning where to build roads or tunnels in order to model the steepness of slopes very accurately.

CHAPTER 5 **Understanding the GIS Data Models** 83

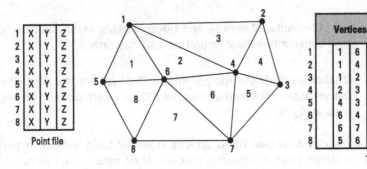

FIGURE 5-6: A Triangulated Irregular Network (TIN) model.

© John Wiley & Sons, Inc.

The TIN model has several benefits:

» **Built-in topology:** Similarly to how intersections connect streets, TIN triangles connect to model elevation smoothly.

» **Flexible conversion:** Need a different perspective? No problem. You can easily transform TINs into digital elevation models (DEMs) or contour lines.

» **Interpolation support:** Got missing data points? TINs use interpolation techniques to estimate unknown values, ensuring smoother and more continuous surface models.

» **Advanced analysis:** TINs let you analyze surfaces in various ways, such as by creating cross-sectional profiles to understand how steep a hill is or modeling potential water flow paths for hydrological studies.

Choosing the right model for surfaces

Both raster and vector models have their strengths. Raster models (like DEMs) work well when you need an overview, like assessing how the terrain changes across a wide region. If you need pinpoint accuracy, such as when determining the exact slope of a hillside, a vector-based model like TIN is your best choice.

Think of raster models as the wide-angle view: perfect for capturing the big picture, like looking over a valley. TINs, on the other hand, are like zooming in with binoculars: ideal for understanding the fine details of that steep hill you're about to climb.

Many GIS projects use both models. You might start with a DEM to create a rough 3D model of a landscape, and then switch to a TIN for a precise analysis of specific features, like a hill slope or a pipeline route.

IN THIS CHAPTER

» Understanding the power of data attributes

» Working with database management systems

» Using object-oriented systems to provide more geographic reality

Chapter **6**

Keeping Track of Attribute Data

GIS isn't just about creating maps — it's also about the data behind those maps. At the heart of GIS are graphics such as points, lines, and polygons, but the attributes connected to these graphics are what provide meaning and context. Whether the attribute is the name of a road, the species of a tree, or the status of a land parcel, an attribute turns simple shapes into useful data for analysis and decision-making.

In this chapter, I walk you through the different ways GIS software platforms manage these attributes, ranging from the use of traditional relational databases to powerful object-oriented systems. You discover how to organize data efficiently, avoid duplication, and tap into new database technologies. I also introduce key concepts like data interoperability, standards, and security to keep your GIS projects running smoothly in today's connected world.

Knowing the Importance of Attribute Data in GIS

Attribute data are what make GIS powerful. Attributes are the detailed information that turns a basic map into a meaningful tool for analysis. Lacking these details is like having a contact list full of names but no phone numbers or emails. Attributes in GIS enable you to ask questions like "Where are the parks with shaded playgrounds?" or "Which areas are at risk for flooding based on land type?"

Whenever you click or tap a map in an app for information like restaurant hours or customer reviews, you're accessing the attributes for that location. You may not have realized it, but by using simple features like that, you've already taken advantage of attribute data!

Exploring the evolution of geospatial data

Attribute data have always been central to GIS. Early GIS platforms focused primarily on mapping and spatial visualization. But over the past few decades the rapid advancement of technology has changed what you can do with GIS.

These are some of the big technological advancements that have put GIS on the map (pun intended):

- **GPS technology:** Provides precise location data, revolutionizing navigation and spatial analysis in GIS
- **Satellite imagery:** Offers high-resolution data essential to monitoring environmental changes, land use, and urban development
- **Cloud Computing:** Enables real-time data access and collaboration, making GIS more scalable and accessible
- **Artificial intelligence (AI) and machine learning:** Automates pattern detection, classification, and predictive analysis within spatial datasets, allowing for faster and deeper analyses of critical data
- **IoT (Internet of Things):** Collects real-time data from sensors, enhancing applications like smart cities and environmental monitoring

Today, advancements in technologies like satellite imagery, cloud computing, and AI mean that GIS is no longer just about *where* things are but also *what* they are and *how* those things change over time. Evolving from simple mapping to dynamic, data-rich systems, geospatial data provide the power behind GIS as a decision-making tool.

Understanding the role of attribute data in GIS

Attribute data bring meaning to spatial features by providing descriptive information that transforms them into more than just points, lines, or polygons. Without attributes, a point on a map is just a spot. But when the point is enriched with attributes, such as the species, height, and age of a tree, that spot becomes a point containing valuable information for urban planners, ecologists, or researchers.

Attributes empower users to ask and answer complex questions. For example, attributes help you filter out which parks have shaded playgrounds, or which buildings fall within a floodplain. This contextual information turns GIS into a decision-making tool that goes beyond simple location mapping.

Attributes fuel spatial analysis. Emergency responders use attributes to locate hospitals with available beds, whereas conservationists use them to monitor changes in wildlife habitats. By adding descriptive context, attributes transform GIS from a simple map into a tool that drives smart decision-making.

Ultimately, attribute data underpin every GIS function, allowing for data comparisons, predictive modeling, and storytelling and helping users understand spatial relationships and communicate findings effectively. For example, analyzing the combination of population, land use, and income for a city over time reveals a story of urban growth.

REMEMBER

With attribute data, spatial features become more than shapes; they become stories, analyses, and insights. Whether you're tracking population growth, monitoring wildlife habitat, or optimizing delivery routes, attribute data ensure that spatial features carry the meaning and insight necessary to inform real-world decisions.

Working with Tables and Database Management Systems

A *database management system (DBMS)* is an essential tool for organizing, storing, and retrieving large datasets. Although traditional DBMS platforms were initially developed for nonspatial data, they have evolved to seamlessly integrate with GIS. Modern GIS software uses a relational database (RDBMS, explained in the next section) and a geospatial database, which is a database specifically designed to store and manage spatial data, to store spatial features (points, lines, and polygons) along with their descriptive attributes. Integrating GIS software and these databases ensures that your data stays organized and ready for smooth updates and analysis.

Today, geospatial databases, such as PostGIS, SQLite/GeoPackage, and Esri's geodatabase, offer enhanced performance and advanced features, like network analysis and spatial queries. (I tell you more about spatial queries later in this chapter, and about network analysis in Chapter 14).

Structuring relational data

At the heart of GIS databases are *relational tables,* which organize data into rows and columns. Each row (or record) represents an individual feature such as a parcel, and each column (attribute) stores a specific type of information, such as land use category or owner, as shown in Figure 6-1.

FIGURE 6-1: A relational database management system (RDBMS) table showing features and attributes.

Attribute (field)

Parcel_ID	Lucode	Status	Region	Zoning
APN12389	1002	active	NW	Agriculture
APN24897	2001	dormant	NW	Agriculture
APN34856	3001	active	SW	Commercial
APN95827	1002	active	SW	Grazing District
APN75394	1004	active	SW	Agriculture
APN35988	1002	dormant	785	Agriculture
APN33580	2001	active	704	Residential

Feature (record)

© John Wiley & Sons, Inc.

A *relational database* gets its name from how it links data between tables using *primary keys* (unique identifiers) and *foreign keys* (references to primary keys in related tables). These keys create relationships between tables, allowing them to share information without duplicating data. This sharing of information is why you often see primary and foreign keys referred to as "shared keys."

Primary keys are often single-column values, like a parcel identification number (such as Parcel_ID in Figure 6-1). However, you can create *composite keys* by combining multiple columns to ensure uniqueness, like combining an owner's name with a street address. Databases automatically index primary keys, making searches faster by directing queries to the right data. The primary key's index acts like a book's table of contents, helping the database quickly find the matching row instead of scanning the entire table. This capability is especially helpful when you have a large dataset.

Take a look at Figure 6-2. The Land Use table lists parcels with attributes like Parcel_ID, Status, and Acres. Each row represents a different parcel, and the Parcel_ID ensures that each entry is unique.

Below the Land Use table is a Transactions table that logs sales and leases using a matching Parcel_ID as a foreign key. Because properties can be bought and sold multiple times, the foreign key is not required to be unique. Instead, it links each transaction to the relevant parcel.

Primary key

Land Use

Parcel_ID	Lucode	Status	Region	Zoning	Acres
APN12389	1002	active	NW	Agriculture	20
APN24897	2001	dormant	NW	Agriculture	40
APN34856	3001	active	SW	Commercial	250
APN95827	1002	active	SW	Grazing District	650
APN75394	1004	active	SW	Agriculture	40
APN35988	1002	dormant	785	Agriculture	250
APN33580	2001	active	704	Residential	80

Foreign key

Transactions

TransID	Parcel_ID	TransDate	Type	Amount
320107	APN12389	2024-06-18	Lease	15000
350718	APN75394	2023-11-06	Lease	18000
640504	APN12389	2014-08-24	Sale	300000

FIGURE 6-2: Primary and Foreign keys in database tables.

© John Wiley & Sons, Inc.

When setting up your main table, every row should represent a unique feature. Use primary keys to ensure uniqueness and foreign keys to link related data between tables.

A good rule of thumb for building a database is to make your tables simple and focused on a specific theme, like land ownership or zoning. Plan ahead by asking yourself, "What questions do I want this table to answer?"

CHAPTER 6 **Keeping Track of Attribute Data** 89

Connecting data with joins and relationships

In GIS, relationships between datasets allow you to link tables without duplicating data. Whether you're combining property records with sales data or linking a land use table to zoning types, joins and relationships ensure that data stays consistent and easy to manage.

How joins work in GIS

Joins connect two or more tables based on shared keys, such as by matching a Parcel_ID from the Land Use table with the same ID in the Transactions table. Using joins, you can access and analyze data from different sources as if they were part of a single table, without duplicating information. Joins are especially useful when data changes over time, such as for tracking property sales or zoning changes.

GIS platforms support several types of joins, with each type being useful for different situations:

>> **Inner join:** Returns only matching records between two tables. This is the most common type of join, useful when you need features that exist in both tables.

>> **Left (or outer) join:** Keeps all records from the first (left) table and only matching records from the second (right) table. When no match is found in the right table, the corresponding columns from the right table are filled with null values, while the left table's data remains unchanged. This type of join is helpful when you want to keep all features from one source, even if no matching data exist in the second table.

>> **Spatial join:** Links datasets based on their geographic relationships, such as by finding which parcels fall within a specific county or which properties are in a flood zone.

Defining relationships between tables

In addition to joins, GIS software uses relationships to link datasets. (In ArcGIS, these relationships are called "relates"; most other software refers to them as "relationships"). Unlike what happens with a one-time join, relationships define how tables interact over time, ensuring consistency even when data is added or updated across related tables.

GIS supports different types of relationships:

» **One-to-one:** Each feature in one table matches a single feature in another. For example, one parcel links to a single owner.

» **One-to-many:** A feature in one table relates to multiple features in another. For example, a single parcel might have multiple zoning changes or sales transactions.

» **Many-to-many:** Multiple features in one table connect with multiple entries in another. For instance, several utility lines might serve multiple properties.

Relationships ensure that updates in one table, like a change in ownership, automatically reflect across linked tables, eliminating redundancy.

Using joins and relationships keeps your data consistent and manageable. Instead of duplicating data in multiple tables, you link relevant tables, making updates easier and ensuring that data stay accurate.

Managing attributes with spatial databases

GIS software — ArcGIS and QGIS, for example — stores vector attributes in spatial database formats such as geodatabases and GeoPackages. These databases combine spatial features with attributes, storing everything in one place and allowing for easy updates and fast queries.

A *spatial database* is more than a spreadsheet; it links features like parcels or roads directly to their attributes, making searches and updates efficient.

Take a look at Figure 6-3, which shows a parcel selected from a land use geodatabase. The attribute table highlights the information for the selected parcel, providing details like its development status (Developed), sector name (Multi Family Residential), and size. Fields like OBJECTID and Shape uniquely identify the geometry of each polygon and connect it to its attributes.

When working with attribute data, you should always ensure data integrity (keeping your data accurate and consistent). Geodatabases help enforce data integrity with domains, subtypes, and validation rules, as follows:

» **Domains:** Domains limit the values that a field can have. For example, you can use a domain to limit the acceptable values for your Zoning field to Agriculture, Commercial, or Residential. Domains are not only helpful for ensuring consistent field values but also avoiding dreaded typos (which can wreak all kinds of havoc when running summary statistics).

CHAPTER 6 **Keeping Track of Attribute Data** 91

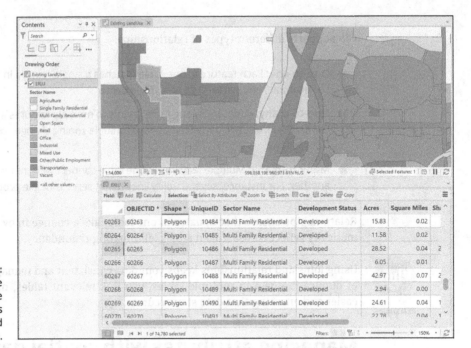

FIGURE 6-3:
A land use geodatabase layer with its associated attribute table.

» **Subtypes:** These are a subset of features that share the same attributes. For instance, you can categorize a road layer by subtypes like local streets, county roads, and highways. Subtypes are primarily used in Esri products, like ArcGIS Pro. In QGIS or other non-Esri software, you can achieve similar functionality using domains and filtering based on field values.

» **Validation Rules:** These rules prevent incorrect data entry. For example, you might include a constraint rule that allows only positive values to be entered for parcel size.

REMEMBER

Using data integrity tools helps make your GIS data more reliable and easier to analyze. Esri's geodatabase format supports tools such as domains for limiting field values, subtypes for categorizing features, and validation rules to prevent invalid entries. Other spatial database types include similar data integrity tools, such as PostGIS/PostgreSQL's CHECK constraints (similar to domains) and validation rules using triggers and constraints.

Exploring pixel-level attributes

In Chapter 5, I explain how vector and raster data models work. When dealing with vector data, you rely on attribute tables, as I describe in the previous sections, and each feature has a unique ID. But raster data doesn't work that way.

92 PART 2 **Geography Goes Digital**

Instead of individual features, raster data are made of a grid of pixels, and each pixel stores a value that represents something, like elevation, land cover type, or temperature. This makes raster data very useful for capturing and analyzing continuous data across large areas, making them ideal for environmental modeling and terrain analysis.

Modern GIS software supports *multi-band rasters*, which store several layers of information within one database. For example, a single satellite image might include multiple spectral bands that detect visible light, infrared, and near-infrared. Scientists often use the red and near-infrared bands to assess vegetation health.

Beyond multi-band rasters, some rasters are multidimensional, meaning they capture changes over time or across depths. This multidimensional feature makes rasters ideal for tracking things like temperature shifts, pollution levels, or water-quality trends over time—great for climate change studies and long-term environmental monitoring.

REMEMBER

Multi-band rasters let you pack multiple data layers into a single file, making them very efficient for analysis. Each band acts like a different page in a book, giving you new ways to explore the same area of interest. Go to https://pro.arcgis.com/en/pro-app/latest/help/data/imagery/raster-bands-pro-.htm to see what a multi-band raster looks like.

Just as with vector data, you can search and analyze raster data. In fact, querying raster attributes is similar to working with relational database tables. Each pixel holds a value for a specific attribute, and GIS software organizes these values into searchable tables. For example, you might search for all pixels with an elevation above 1,000 meters, or identify forested areas based on pixel values.

Raster queries often focus more on patterns and combining multiple layers than on individual values. Because each raster cell contains one value per layer, overlaying different raster datasets (like land cover, slope, and temperature) allows you to compare attributes across the location and answer complex spatial questions:

>> Which areas of the forest have steep slopes and temperatures above 80° Fahrenheit?

>> Where are the best sites for wolf habitat?

Whereas vector data is great for representing discrete objects like roads and parcels, raster data is better suited for modeling continuous surfaces, like terrain elevation and temperature maps. In many GIS projects, you'll find your work relying on both raster and vector data to answer the tough questions.

CHAPTER 6 **Keeping Track of Attribute Data** 93

For example, you might use vector polygons to represent watershed boundaries while overlaying them with a raster elevation layer to analyze drainage patterns. Or you might combine land use polygons with a raster slope map to find areas where development might be challenging. For more details on getting information out of rasters, check out Chapter 8.

Using both raster and vector data in your GIS project allows you to see the big picture and dive into details. Vector data gives you precise locations, and raster data fills in the environmental details to provide more context.

Searching with SQL in GIS

Modern GIS software relies heavily on Structured Query Language (SQL) to retrieve and manipulate data stored in spatial databases. Whether you're working with vector or raster data, SQL offers powerful tools for filtering, combining, and analyzing data.

You hear SQL pronounced two different ways: Some folks spell it out as "S-Q-L," and others say "SEE-quell." Both are correct, so don't be surprised if you hear it both ways in GIS and database discussions.

SQL in GIS allows you to structure queries using command words like these:

>> **Select:** Extract specific attributes from a table, as in this example:

```
SELECT Owner, Land_Use FROM Parcels
```

>> **Where:** Add conditions to fine-tune your query; for example:

```
WHERE Acres > 80.
```

>> **Join:** Combine information from multiple tables based on common keys.

```
Example: JOIN Zoning on Parcels.Zone_ID = Zoning.Zone_ID
```

>> **Order by:** Sort your results, as in the following:

```
ORDER BY Parcel_ID.
```

SQL is a standard computer language designed specifically for accessing, querying, and manipulating databases. It allows you to search for very specific information, retrieve it, delete or add to it, update it, and combine queries. SQL searches database tables by using commands such as SELECT, AND, OR, INSERT, DELETE, ORDER BY, and many more.

Here's a simple example of how SQL can help you search GIS data. Imagine that you have a table of land parcels named Land_Use with fields (attributes) named Parcel_ID, Lucode, Status, Region, and Zoning (Refer to Figure 6-1).

If you need only the Parcel_ID and Zoning for each feature, you can write a query like this:

```
SELECT Parcel_ID, Zoning FROM Land_Use;
```

This query pulls the Parcel_ID and Zoning attributes from the table, returning a smaller, more focused dataset that looks something like this:

Parcel_ID	Zoning
APN12389	Agriculture
APN24897	Agriculture
APN34856	Commercial

SQL makes it easy to extract exactly the information you need from large datasets, whether you're analyzing property data or querying metadata from raster datasets (like spatial resolution or data type). As you build more complex queries, you can combine multiple commands to refine your analysis.

SQL is a valuable tool to have in your GIS toolbox. Most GIS software, like ArcGIS, QGIS, or GRASS, comes with built-in tools or plug-ins that enable you to use SQL to work with spatial data. Understanding SQL is a huge asset for any GIS user. Check out Chapter 9 for more on using SQL, including some advanced SQL functions.

Understanding Object-Oriented and NoSQL Systems

An *object-oriented database management system (OODBMS)* extends the traditional database management system by storing data as objects. Each *object* represents a collection of data with built-in rules and behaviors, making it smarter than a standard row in a relational database.

Objects have two primary characteristics that make them powerful for GIS:

» **Inheritance:** Objects share properties with their parent types. For example, in a transportation infrastructure system, every road (a "child" object) inherits general properties like "surface type" from its parent object.

» **Hierarchy:** Objects exist within a structure, such as roads categorized as highways, local roads, or bike paths, each with unique rules for how they interact.

In GIS, object-oriented principles prevent errors and ensure data integrity by enforcing rules. For example, a sewage system model won't let you connect incompatible data, such as putting two-inch pipe and eight-inch pipe together, thereby helping to prevent design mistakes. This rule enforcement gives you a huge advantage when working on large, complex projects such as infrastructure management or environmental monitoring.

Relational, object-oriented, and NoSQL databases each handle data differently, making them suited for different GIS applications. See Table 6-1 for a quick comparison of how these database types are structured and used in GIS. (For a refresher on how GIS enforces rules, see the section "Adding spatial smarts to GIS with topology" in Chapter 5.)

REMEMBER

Object-oriented databases enable GIS tools to share rules and behaviors directly in the data, helping you avoid errors and ensure consistency.

TABLE 6-1 Comparing Database Types

Feature	Relational (SQL)	Object-Oriented	NoSQL
How data are stored	Data are organized in tables with rows and columns.	Data are stored as objects with properties and relationships.	Data are stored in a flexible, nontabular format, such as document-based storage (JSON/XML).
How relationships work	Uses primary keys (unique IDs) and foreign keys to link tables.	Objects inherit properties from parent types.	Does not rely on traditional table relationships; stores loosely connected data that scales efficiently.
How it works in GIS	GIS features are stored in related tables to avoid duplication.	GIS features behave as objects with rules and hierarchy.	GIS can handle massive, real-time datasets from sensors, social media, or cloud platforms.
Best for	Structured data with clear relationships.	Complex GIS models where features interact.	Big Data and Internet of Things (IoT) applications.

Enhancing descriptive information with object orientation

On the surface, a *geodatabase* (Esri's implementation of OODBMS) looks similar to any relational database. It has tables with columns and rows, and it uses primary and foreign keys to connect tables. But underneath the surface, the geodatabase goes deeper by organizing data into objects with their own set of rules and behaviors.

For example, take a look at infrastructure models:

» A road network object can store attributes like road type and surface material while also containing rules about speed limits and one-way streets.

» A water utility object might store the pipe diameter along with rules to ensure compatibility between connected pipes.

These rules help maintain data integrity by not allowing entries that don't meet the established rules for the object's class. If you try to connect a two-way street with a one-way street going against the flow, your GIS will stop you from making a wrong move.

TIP

Think of object-oriented GIS as giving your data "street smarts." Not only does each object store descriptive information, but — with the inclusion of behaviors and relationships — it also knows how it can behave and interact with other objects in the system.

Exploring emerging data storage systems

As GIS continues to expand into new areas like smart cities, the Internet of Things (IoT), and big data, traditional relational databases are no longer enough. The need to handle unstructured data — large, rapidly changing datasets that don't fit neatly into rows and columns — is here. That's where NoSQL database management systems come in.

NoSQL (a.k.a. "not only SQL") databases, like MongoDB or Cassandra, are typically used in applications that have huge amounts of data requiring real-time processing and analysis, like traffic flows, air quality sensor readings, and social media streams. As the name implies, NoSQL databases typically do not use SQL to interact with data and do not rely on relationships (they're non-relational). Instead, NoSQL allows you to store data in flexible formats with no fixed schema, making it easier to manage large-scale and diverse data.

Cloud storage has also revolutionized GIS, allowing for massive datasets to be accessed, stored, and managed in real-time across the globe. Cloud-based platforms like Google Cloud, Microsoft Azure, and Amazon Web Services (AWS) allow you to store vast amounts of accessible, on-demand data. These systems have proven beneficial in supporting applications in environmental monitoring, smart cities, and more.

TIP

Consider the use of NoSQL in your GIS in the following situations:

- » **Handling large datasets in real-time:** NoSQL is a good choice if you're managing continuously generated spatial data from IoT sensors, such as traffic flow or air quality.

- » **Storing diverse data types:** NoSQL supports both structured and unstructured data, making it useful for integrating geospatial data with sources like social media feeds or sensor data streams.

- » **Scaling with cloud and big data:** NoSQL works well with platforms like Google Cloud or AWS, allowing you to process large datasets and scale up without overhauling your entire system.

Leveraging geodatabases and industry models for success

GIS is a powerful tool used in an enormous number and variety of industries. However, by trying to meet the needs of all these different users, GIS might be seen as too large and complex. This is where leveraging the geodatabase model pays off. The geodatabase model is widely used across industries and organizations because it allows you to create custom data models (schema) tailored to your industry. Luckily, to help you avoid starting from scratch, Esri offers industry-specific geodatabase models. For example:

- » Transportation models help manage road networks, including rules for traffic flow and connection.

- » Water utility models store pipe networks, with rules ensuring that pipes connect only at valid points.

- » Environmental models track wildlife habitats or monitor changes over time, using multidimensional data.

These geodatabase models often include

>> **Case studies** to show how the model works in real-world situations

>> **Templates** for you to import your own data

>> **A poster or diagram** that visually explains how all the data fit together

>> **Tools** for customizing the model to your specific needs

TIP

Using industry-specific geodatabase models saves time and ensures that your data meets industry standards. Also, you don't have to build everything from scratch; you can just customize the model to fit your project!

Incorporating Data Interoperability, Standards, and Security

As GIS continues to evolve and integrate with more technologies, ensuring interoperability, adherence to standards, and data security becomes essential. With the rapid rise of cloud computing, artificial intelligence (AI), and machine learning (ML), GIS systems must be able to exchange data seamlessly, maintain accuracy, and protect sensitive information.

Integrating AI and machine learning in GIS

AI and ML are transforming GIS by automating data classification, identifying patterns, and predicting trends, giving GIS users faster insights. For example, ML methods use algorithms to create land-cover classifications, predict crime hotspots, and forecast the spread of disease.

TIP

With AI, you can build predictive models for urban development and automate time-consuming tasks like mapping vegetation health. Seamlessly integrating GIS data with AI tools enables you to make data-driven decisions and unlock deeper insights faster than ever before.

Ensuring interoperability with standards

Data interoperability lets systems and sources communicate, ensuring that they work together seamlessly. Following standards like those from the Open Geospatial Consortium (OGC) ensures consistency and collaboration. OGC standard formats like GeoPackage, GeoJSON, and 3D Tiles work seamlessly in ArcGIS, QGIS, and other tools, making it easy for you to share and collaborate on projects with a variety of users. Interoperability standards are especially critical for projects involving multiple organizations, such as in environmental monitoring programs and smart-city initiatives. Without standards, you may as well be working inside a black box.

Following standards makes collaboration easier and prevents compatibility headaches when working with external partners or publishing data online.

Addressing security and privacy in GIS

As more GIS data moves to the cloud, securing attribute data and protecting sensitive information is essential. Cloud-based platforms like Google Cloud, Microsoft Azure, and Amazon Web Services (AWS) provide powerful, scalable storage systems that allow real-time access. But it's up to you to protect sensitive information. Follow these best practices for keeping your GIS secure:

» **Data encryption:** Ensure that your data are encrypted both within your system and while in transit to another.

» **Configure HTTPS:** Configure your website to use HTTPS instead of HTTP to encrypt connections between your site and visitors.

» **Control access:** Use role-based permissions to ensure that only authorized users (those in an assigned role) can access sensitive information.

» **Anonymize data:** Remove personally identifiable information (PII) when sharing datasets.

» **Compliance:** Follow regulatory frameworks such as the General Data Protection Regulation (GDPR) or ISO standards for data protection.

» **Educate:** Stay informed about key principles and best practices for data security, protection, and governance, and ensure your team understands them, too.

Take security seriously: Keep your systems updated, train your team on best practices, and regularly review your data access policies.

> **IN THIS CHAPTER**
> » Using the best data for your GIS project
> » Incorporating data from GPS, remote sensing, and online data sources
> » Documenting your data

Chapter 7
Collecting Geographic Data

Every GIS project relies on quality data in order to produce meaningful and accurate results. Whether you're mapping customer locations, tracking environmental changes, or analyzing urban infrastructure, the integrity and relevance of your data determines the power of your GIS.

In this chapter, I guide you through the essential steps to ensure that your data is ready for use in GIS. You start by exploring what it takes to get high-quality data and the factors to consider when identifying reliable sources. From there, I show you the various ways to bring data into your GIS, such as by importing it from external sources, geocoding addresses into mappable points, and fetching real-time information through API connections. Last but certainly not least, I explain why metadata — the data about your data — is critical for understanding, sharing, and maintaining all your geospatial data.

Identifying Quality Data

Good data are the bedrock of any GIS project. Without data, even the most advanced maps or tools won't give you reliable results. The sources of your data — along with how accurate, timely, and complete they are — can make or break your analysis. GIS data are like the nutrients in a garden; with the right nutrients, plants grow strong and healthy. Likewise, good data provide nourishment for your GIS project, allowing it to flourish and yield valuable insights. High-quality data ensure that your GIS work supports good decisions and helps you avoid costly mistakes.

Understanding why quality matters

Say you're mapping wildlife habitats as part of a conservation project. If all your animal sightings come from a single hiking trail (not because the animals prefer that area, but because more observers collect data there), the range of those animals won't be accurately reflected on your map. Or imagine using outdated aerial imagery to plan new bike paths through a city. If the region has seen a lot of recent development, your plan could lead cyclists into closed-off areas or newly built neighborhoods.

REMEMBER

Getting data that's timely, accurate, and appropriate ensures that your GIS does what it's supposed to do: guide decisions that you can trust.

Evaluating factors for collecting high-quality data

Here are some key questions to ask yourself as you gather data:

» **Is the scale or resolution appropriate?** How much detail do you need? If you're mapping tree clusters in a forest, a general map showing only the forest outline won't cut it. You need a detailed map showing individual trees and clusters so that you can see the arrangement and density of the trees in the forest.

» **Do you need categories or precise numbers?** Some analyses require only categories, such as land use classifications (urban, forest, agriculture). Others need exact measurements, such as air temperature or elevation. Think about whether your analysis needs general descriptions or specific, measurable data to answer your question.

» **How accurate are the data?** Accuracy matters. A small error in elevation data can make a big difference in flood modeling. The more accurate your data, the more reliable your analysis.

» **Are you collecting in the right places?** If you're sampling water quality, for example, are you testing in locations that represent the entire watershed? Skewed sampling leads to biased results.

- » **How fresh is the data?** Some analyses require real-time or up-to-date data. Outdated information — like using old zoning maps to plan new infrastructure — can lead to bad decisions.

- » **Do you have the right data type and format?** If you're mapping tree cover, do you need field measurements, drone imagery, or satellite data? Make sure your data types fit your project's goals and that formats are compatible with your GIS software.

- » **Do the data classifications align?** Inconsistent classifications can mess up your analysis. For example, combining land-use classes from different years with mismatched categories can make it hard to compare results.

- » **Is the dataset complete?** Do you have everything you need to answer your question? Missing data, such as incomplete traffic counts for a transportation study, can throw off your analysis and conclusions.

TIP

Collect only the data you need. It's tempting to gather more, but too much irrelevant information just makes your project more expensive and harder to manage. Stay focused on what matters.

Importing Data

Data can come from a variety of sources, each with its own strengths and weaknesses. The method you choose and type of data you collect impacts the insights you can gain, whether you're collecting bird sightings with GPS, analyzing aerial images, or accessing data online. In this section, I introduce you to some common ways to bring data into your GIS, from GPS and remote sensing to sampling techniques. Understanding these methods can help you choose the right data sources and set up your data for effective analysis in your GIS projects.

Collecting data with a GPS receiver

Almost everyone seems to use a GPS these days, whether it's through Google Maps, Apple Maps, Waze, or popular activity-tracking apps like AllTrails, or Strava. These apps have become so common that often when I tell people I work in GIS, they say, "Oh, cool; I use GPS all the time on my phone!" Well, they're sort of similar, but not quite! A GIS (Geographic Information System) and GPS (Global Positioning System) are related, but they're different tools.

Your smartphone, for instance, is a GPS *receiver*, which means that it can determine your location using signals from satellites. But unlike dedicated GPS devices, most smartphones don't rely solely on satellites. Instead, they use *assisted GPS* (A-GPS) to speed up location tracking and improve accuracy. A-GPS combines satellite signals with data from cell towers and Wi-Fi networks so that your phone can determine your position when GPS signals are weak, like indoors, in urban areas with tall buildings, or under tree cover. A-GPS technology is why your phone still shows an approximate location even when it doesn't have a clear view of the sky.

GPS satellites are part of a broader system called Global Navigation Satellite System (GNSS), which includes not just GPS but also satellites from other countries, like Europe's Galileo or Russia's GLONASS. Together, these GNSS satellites help provide even more reliable location data, especially in places with limited satellite visibility. Although people often use the term *GPS* when they really mean *GNSS*, I stick with GPS here to keep things simple.

For GPS to work, it relies on a network of more than 30 satellites orbiting the Earth, along with a network of ground-based stations (like the one shown in Figure 7-1) and a receiving device (such as your smartphone or a handheld GPS like the one in Figure 7-2). By measuring signals sent from multiple satellites, GPS receivers can determine your precise latitude (north-south), longitude (east-west), and even elevation.

FIGURE 7-1: A typical GNSS base station.

Philipp Berezhnoy/Adobe Stock Photos

FIGURE 7-2:
A typical handheld GPS device.

sanderstock/Adobe Stock Photos

For best accuracy, a GPS receiver needs a clear line of sight to at least four satellites, which is why some of your favorite phone apps lose track of you along tree-covered trails or as you're weaving through the grocery store aisles. However, because modern smartphones also use Wi-Fi and cell towers for positioning, they can often estimate your location even in areas where satellite signals are weak. In open areas, even basic GPS units can be impressively accurate, making them reliable for general GIS data collection when a good satellite connection exists.

With advances in both smartphone apps and dedicated GPS devices, GIS field data collection is accessible and versatile. You can collect data on invasive plants, locate and count fire hydrants in your city, or track your hike along a segment of the Continental Divide Trail — all with your smartphone, tablet, or handheld GPS Device.

TECHNICAL STUFF

Smartphones typically achieve around three to ten meters (10 to 30 feet) of accuracy, which is often enough for general data collection like tracking locations or routes. However, for professional GIS work that requires higher accuracy, pair your smartphone with an external GPS/GNSS receiver or use a rugged handheld GPS device designed for field work.

TIP

Handheld GPS devices, like the one shown previously in Figure 7-2, are widely used for general field mapping and navigation. They offer durability, good accuracy, and can store points, tracks, and routes that are easily exportable to GIS. High-end devices used for environmental monitoring, infrastructure mapping, and professional land surveying can achieve sub-meter and even centimeter-level accuracy. If your work requires that level of accuracy, however, it's best to leave the data collection work to the professionals.

Using remote sensing to collect data

Have you ever spent time looking out the window of an airplane in search of familiar landmarks and features below? That's essentially what remote sensing is all about: capturing images of the Earth from above and using them to map everything from natural landscapes to urban areas. Scientists have been using images this way for well over a century, beginning with aerial photos used in World War I for military reconnaissance and mapping. Today, satellites, drones, and other digital technology make it easy to gather fresh data regularly.

TECHNICAL STUFF

When satellites capture images of the Earth, they collect data in raster format — a grid pattern, in which each small square, or pixel, represents a specific area on the ground (I describe raster data in Chapter 5). Remote sensing relies on different types of light (or "bands") covering various parts of the electromagnetic spectrum.

Our eyes can see only within the visible light range, but remote sensors pick up a much wider range, including *infrared* (helpful for analyzing vegetation and soil moisture) and *microwave* (used in radar sensing, which can see through the clouds). (Go to https://www.earthdata.nasa.gov/learn/earth-observation-data-basics/remote-sensing for more about the spectrum bands that remote sensors use.) By combining different bands, remote sensing provides a more complete picture of the environment, making it a powerful tool for mapping and monitoring changes on the Earth's surface.

Comparing passive versus active remote sensing

The key difference in remote sensing is whether the sensor relies on external energy sources or generates its own signal to gather data. These are known as passive and active sensors and work like this:

- » **A passive sensor detects naturally occurring energy, such as sunlight or heat.** When the sun's energy bounces off surfaces, passive sensors pick up the reflected light, which is great for daytime imaging. It's like taking a picture without using a flash.

- » **An active sensor emits its own signal and measures how that signal reflects off objects.** Technologies like radar and LiDAR (Light Detection and Ranging) use this method to collect data even in low-light conditions, such as at night, in dense forests, or beneath the ocean's surface. An active sensor works like a camera using a flash—it sends out light that bounces off a subject and back to the camera's sensor.

106 PART 2 Geography Goes Digital

Enhancing and classifying images

Ever taken a photo, noticed that the lighting was off, and thought, "I'll just edit that later?" Well, remotely-sensed images often need a bit of photo editing, too. This process, called *image enhancement*, makes details clearer and easier to interpret, helping you spot boundaries or other fine features in your imagery.

You might use image enhancement to pull out hidden details in satellite imagery. Tools like contrast stretching make bright areas brighter and deepen shadows, and sharpening tools bring out fine edges. Together, these tools turn a good image into one that's clearer, more informative, and ready for mapping!

Image enhancement is especially helpful when features are hard to distinguish with the naked eye, as in satellite images of urban areas, forests, or coastlines. Enhanced images can reveal key details that support detailed mapping and analysis.

If you want to identify specific features such as bodies of water automatically, you can use classification tools to group similar pixels together. But rather than do this work yourself, you can teach your GIS software to classify the data for you in two different ways:

>> **Supervised classification:** Using this method, you teach the software to recognize features by picking pixels of a known feature, like a lake, and having the computer find other pixels with similar values. This method is ideal when you know what you're looking for and want accurate, specific results.

>> **Unsupervised classification:** With this method, you let the software look for patterns on its own, grouping pixels without your guidance. It's a quick way to reveal natural patterns in areas you're less familiar with. You might use it to discover clusters of land cover types in a new region, for example.

You can load your enhanced and classified images directly into GIS, where they become valuable data layers. For example, you can add a land cover layer to analyze changes in urban growth over time, or add a topographic layer to map out elevation.

Using satellite imagery as part of GIS data provides a fast, efficient way to monitor and update large areas. These data are particularly useful for tracking land cover changes or updating maps because new images are available regularly and at a relatively low cost.

Working with samples of data

People collected data in the field long before satellites, GPS, or even GIS existed. Field data collection might involve house-to-house surveys, counting cars on roads, recording air temperature, or gathering soil, vegetation, and insect samples. For centuries, naturalists, anthropologists, and many others have documented people, plants, animals, landforms, and more. But collecting data with the specific goal of creating maps came a bit later.

Although field data collection has been a part of science for centuries, collecting data with the specific aim of creating maps brings unique challenges. One of the biggest is that, in most cases, you can't capture every single instance of what you're studying. Imagine, for example, that you want to study the grass in your backyard. You'll spend ages documenting each blade of grass! Or suppose you want to measure the temperature across an entire city. Because temperature is continuous — it's everywhere, all the time — collecting every single measurement is impossible. In both cases, you need to collect data from a sample of the whole. In GIS, you're sampling geographic space, which is continuous by nature.

Maps represent space, so the samples you take (spatial sampling) must accurately represent that space. In other words, if you're studying trees on an island, you can't just sample one side of the island. You need to sample wherever trees might be located across the entire island. Lack of spatial sampling is a common issue with many GIS databases and can limit the accuracy of your analysis.

REMEMBER

When you take samples, make sure you collect them from a variety of places — places with different sizes, shapes, elevations, and climates.

By sampling, you're estimating the numbers, distributions, and locations of geographic features across your study area. Whether you're collecting point, line, or area data, several sampling techniques can help you determine where to sample. Common methods, shown in Figure 7-3, include the following:

» **Clustered:** In clustered sampling, you focus on areas with a high density of features. This approach is useful when features naturally occur in groups or when studying "hotspots." For example, you might use clustered sampling to study plant species in areas with dense vegetation.

» **Systematic:** With this method you sample in specific, fixed intervals along a line or across a grid. This method is commonly used for consistent coverage across an area. For example, you can sample water quality every 100 meters along a river to ensure even representation of the entire length.

» **Random:** Random sampling has no set pattern. A computer often generates random points, so human bias doesn't affect the randomness. This method is commonly used when you want each location to have an equal chance of being selected, giving an unbiased picture of your study area.

» **Adaptive:** This technique involves sampling more often in areas with high variability (where things change a lot) and less frequently in more uniform areas. This method helps you capture detail where it matters most while saving time and resources in areas with little variation. For example, if you're studying soil composition, you might take more samples in areas with changing soil types and fewer in areas with uniform soil.

You can also *stratify* your samples, meaning you divide your area into different groups, or *strata*, to ensure balanced sampling in each part. For instance, to sample who watches certain TV shows across your city, you can divide the city into neighborhoods. Then, in each neighborhood, you choose a certain number of people (say, 25) to sample randomly, systematically (for example, every fifth house), or clustered (focusing on high-density areas).

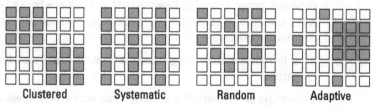

FIGURE 7-3: Common spatial sampling methods.

Clustered   Systematic   Random   Adaptive

coolvectormaker/Adobe Stock

REMEMBER

Spatial sampling helps give you a true representation of your area. Whatever method you use, remember that your data has a spatial pattern. A sampling strategy that ignores that pattern often results in data that don't fully represent the geography.

Geocoding Data

Type a street address into your favorite mapping app (Apple Maps, Waze, Google Maps, MapQuest, or another one) and watch how it zooms into the location and places a dot or icon there. Congratulations — you just geocoded a point! Geocoding is a key GIS skill that allows you to map and analyze all kinds of location data, from customer locations to traffic incidents along highways.

Turning addresses into points

Geocoding converts street addresses and other location details into geographic coordinates, making it easy to map and analyze locations. For example, if you have a list of customer addresses, geocoding can place each one on a map, helping you visualize patterns and plan deliveries.

A geocoding tool, often called a *geocoder*, locates an address on a map by reading the address, searching a database for matches, assigning coordinates, and plotting the point. Here are the basic steps in more detail:

1. **The geocoder reads the address as a complete unit, with known addresses stored as full address strings.**

   In areas without exact address points, the geocoder breaks down the address into parts like house number, street name, city, state, and zip code. Each part helps narrow down the location.

2. **The geocoder then searches a reference database for matches.**

   The database might include data from government sources, commercial providers, or open data like OpenStreetMap. In cases where an exact match isn't available, the geocoder may approximate a location based on the parsed address components.

3. **After it finds a match, the geocoder assigns the address's latitude and longitude.**

   If there's no exact match, the geocoder may assign the address to a less specific location, like the center of the city or zip code.

4. **With the coordinates assigned, your GIS software places the point on the map.**

   The plotted point is ready for you to view and analyze.

Most geocoding tools allow for single address geocoding or batch geocoding when you have a list of addresses to map. Batch geocoding is great when you have a lot of addresses because it places all points on the map simultaneously, saving you time.

TECHNICAL STUFF

Geocoders typically rely on point data, where each address corresponds to a known latitude and longitude, making this method ideal for well-defined addresses in urban and suburban areas. In rural areas, however, where precise address points are often unavailable, addresses are interpolated along a road segment based on address ranges. For example, an address like "1200 Main Street" will be placed halfway between 1100 and 1300 on Main Street. This type of interpolation differs from linear referencing, which uses more precise measurements along calibrated routes (see the next section for more about linear referencing).

Creating points and lines by reference

In GIS, some data points don't have exact coordinates but are located relative to a known point along a route, such as a mile marker on a highway or a specific distance along a river. The process of placing these points on your map is called *linear referencing*. This technique lets you map events or features based on their position along a known line instead of using traditional coordinates. Figure 7-4 shows a simple diagram of how linear referencing works.

Linear referencing is especially useful for data like

- » **Highway crashes:** Instead of giving exact coordinates, a crash report may indicate that the incident happened a quarter mile from an intersection or mile marker. Linear referencing plots the crash location at the correct distance along the route. For example, in Figure 7-4, an incident's position is plotted on a map by measuring its distance from the intersection of Juniper Street and Oak Drive.

- » **Water utility sampling:** If water samples are taken every 100 meters along a river, linear referencing can map these points along the river's flow line.

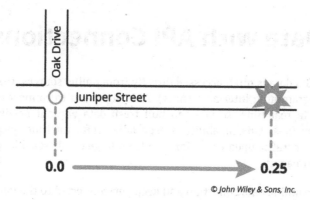

FIGURE 7-4: Plotting an incident on a map.

© John Wiley & Sons, Inc.

To use linear referencing, GIS relies on a *calibrated route feature* (such as a road or river) with measure values. This setup forms a *linear referencing system* (LRS), which includes these core elements:

- » **Calibrated route feature:** A line feature with *M values* (measure values) indicating distances along the route, often measured in miles or kilometers. These values are essential for identifying and locating events along the line.

- » **Reference line:** The specific path along which locations are measured, such as a highway, hiking trail, or pipeline.

- » **Starting point:** Where measurements begin, often marked as 0, or based on a distinct landmark, like an intersection. All other measurements are calculated as distances from this point.
- » **Measure distance:** The distance along the line from the starting point to the location of each event. For example, if a crash occurs 12 miles from the start of the highway, you'll use this measure distance to accurately plot the point.

Here's a quick guide for using LRS to create points or lines:

- » **Check your data:** Ensure that your data includes the route identifier and a measure. For points, you need one measure; for lines, you need both start and end measures.
- » **Map the events or features:** With a calibrated route, your GIS places features along it using the measure values from your data.
- » **Verify placement and accuracy:** Review the placement to ensure that distances are accurate, especially if location accuracy is crucial. This step is essential when working with LRS data that requires alignment with other datasets, such as infrastructure or environmental features.

Accessing Data with API Connections

Today, GIS data is often accessed directly from online sources, thanks to application programming interfaces (APIs). APIs allow you to connect to different data sources in real time, so you can pull fresh data without needing to manually download files. This capability is especially useful for time-sensitive data, like weather or traffic updates, or for data from large databases, like census tables or satellite imagery.

APIs streamline data collection and keep you connected to the latest information, whether you're analyzing population trends, monitoring natural disasters, or mapping out real-time traffic patterns. In this section, you discover what APIs are, how they work, and how to access data using scripting languages like R or Python. For a deeper dive into automating your GIS, check out Chapter 18.

Understanding APIs

APIs act like bridges that allow software systems to connect and communicate with each other. Instead of manually downloading a file, APIs let you request specific data directly from a service provider, such as the U.S. Census Bureau or OpenWeatherMap, and have it pulled directly into your GIS software or script.

Most APIs follow a standard format, so you can specify what type of data you want as well as the time frame, location, or other filters. The API translates your request into something the provider understands, retrieves the data, and sends it back in a structured format (usually JSON or XML).

For GIS users, APIs are invaluable because they offer access to vast, up-to-date datasets for real-time analysis. For example:

- You can use an API to fetch live weather data for emergency response mapping.
- Census APIs let you pull population and demographic data for specific regions directly into your analysis.
- Transportation APIs provide real-time traffic and public transit information that you can integrate into routing and logistics maps.

Fetching data from APIs through scripting

To use APIs in GIS, you typically need to write scripts in a programming language like R or Python. These scripts help you send requests to the API, handle the data it returns, and format the data for use in your GIS software. R and Python both have libraries — such as `httr` in R and `requests` in Python — that make working with APIs easier.

Here's a simple R script that fetches the current temperature for Frisco, Texas from OpenWeatherMap's API:

```
# Load required package
if (!require(httr)) install.packages("httr")
library(httr)

# Set your OpenWeather API key
# get API key at https://openweathermap.org
api_key <- "YOU_API_KEY_HERE"

# Define the location and endpoint
city <- "Frisco"
state <- "TX"
country <- "US"
endpoint <- "https://api.openweathermap.org/data/2.5/weather"
```

```
# Make the API request
response <- GET(endpoint, query = list(
  q = paste(city, state, country, sep = ","),
  appid = api_key,
  units = "imperial" # Use "imperial" for Fahrenheit use "metric" for celsius
))

# Check if the request was successful
if (status_code(response) == 200) {
  # Parse the response content
  weather_data <- content(response, "parsed")
  
  # Extract relevant information
  current_temp <- weather_data$main$temp
  feels_like <- weather_data$main$feels_like
  weather_description <- weather_data$weather[[1]]$description
  
  # Print the data, change printed units to °C if using metric
  cat("Current temperature in", city, ":", current_temp, "°F\n")
  cat("Feels like:", feels_like, "°F\n")
  cat("Weather description:", weather_description, "\n")
} else {
  cat("Failed to retrieve data. Status code:", status_code(response), "\n")
}
```

The preceding script sends a request to OpenWeatherMap's API for the current weather in Frisco. It then extracts and prints the temperature and weather description from the returned data.

When working with APIs, always review the documentation for details on the available parameters and data formats. Many APIs also require an API key for authentication, so make sure to sign up for access and keep your key secure.

To use API data in your GIS software, you may need to convert it to a format like CSV or GeoJSON. R's dplyr package and Python's Pandas library are great tools for manipulating API data before importing into GIS.

Building Data about the Data

After you create a clean, accurate geodatabase, shapefile, or GeoPackage, you might think "I did it! Now I can share this data. It's ready!" Well, not quite. Documenting what your database contains, how accurate it is, how it was collected, and how it should be used is just as important as clean, accurate data.

Metadata, or data about data, serves as a complete description of the key characteristics of a dataset: quality, sources, format, accessibility, and any limitations. This information makes the data not only easier for you to understand and use but also to share, especially with others who may not understand the full background of the dataset.

For example, imagine that you're working on a project involving wetlands. You may have multiple datasets from different agencies, but the definition of "wetland" varies from agency to agency. Without metadata to clarify the definition from each agency, comparing or analyzing the data accurately becomes nearly impossible. Consistent, complete metadata can bridge this gap, providing essential background for users who rely on shared datasets.

TIP

Today, most GIS software includes built-in tools to easily create and manage metadata. Many of these tools adhere to established metadata standards, making compliance much easier.

The Federal Geographic Data Committee (FGDC) is responsible for maintaining spatial data standards in the U.S. government. The FGDC has developed a standard framework for geospatial metadata, the Content Standard for Digital Geospatial Metadata (CSDGM), which includes ten essential elements to ensure that your GIS data can be understood, shared, and reused effectively:

» **Identification:** Describes basic information about the dataset, including its name, creator, coverage area, categories, themes, and access restrictions.

» **Data quality:** Describes the accuracy, completeness, consistency, and reliability of the data. This includes sources for derived data and any known limitations.

» **Spatial data organization:** Provides details about the data format (vector/raster), spatial objects, and any additional location encoding methods.

» **Spatial reference:** Includes the coordinate system, projection, datum, and any specific parameters used for coordinate transformations.

» **Entity and attribute information:** Supplies information on specific attributes, the data each one encodes, codes, and the code definitions.

» **Distribution:** Tells where and how the data can be accessed, including formats, online availability, costs, and the organization that distributes it.

» **Metadata reference:** Includes the date the metadata was compiled, who compiled it, and what metadata standard was used.

» **Citation:** Lists references and sources for the dataset, ensuring proper attribution and traceability.

CHAPTER 7 Collecting Geographic Data   **115**

> **Time Period:** Indicates when the data were collected, which is essential for understanding the dataset's relevance and currency.

> **Contact Information:** Provides details for reaching the data provider, allowing users to ask questions or obtain further information.

In 2010, the FGDC endorsed the international metadata standard ISO 19115 as a transition from CSDGM. ISO 19115, developed by the International Organization for Standardization (ISO), provides a more flexible and internationally recognized structure for describing geospatial data. Many organizations have since adopted ISO 19115 to ensure compatibility across different systems and countries.

Many organizations, including government agencies, provide online tools and templates to help you create metadata. For example, the U.S. Geological Survey offers metadata guidelines and examples through its online resources. Most modern software like ArcGIS and QGIS includes integrated metadata tools that allow you to create, edit, and link metadata to datasets directly within your GIS project, often supporting both CSDGM and ISO 19115 formats.

WARNING

Always include current, complete, and accurate metadata for your datasets, and be cautious about using data from sources that don't provide it. Accurate metadata ensures that anyone using your data understands the quality, limitations, and appropriate applications.

If your GIS software doesn't include a metadata tool, you can explore third-party tools specifically designed for metadata creation and management. These tools can be particularly useful for batch processing, validation and compliance, and ensuring alignment with specialized metadata standards.

MAPPING NEARBY PARKS FOR YOUR NEIGHBORHOOD

The best way to familiarize yourself with data is to simply start working with it. Say you want to map all the parks within five or ten miles of your home. Here's how you can do it using data from OpenStreetMap (OSM) and QGIS, the free, open-source GIS software:

1. **Go to the HOT Export Tool** (https://export.hotosm.org/) **to download parks in your area of interest.**

    You can use either a GeoPackage, Shapefile, or Geojson format. (Search for "parks" in the Tag Tree.)

2. **To load the data into QGIS, open QGIS and add the downloaded park data as a layer.**

   Don't worry if it seems complicated; QGIS has step-by-step tutorials to get you started. (Try loading the OSM basemap layer from the XYZ Tiles section in the QGIS Browser Panel).

3. **Zoom to your neighborhood and add a point.**

   Use the zoom and pan tools in QGIS to focus on your home and surrounding area. Create a new point layer and add a point where your home is located. QGIS has an option to create a temporary scratch layer, which is helpful if you won't be needing the layer again later.

4. **Highlight nearby parks.**

   Use the "Buffer" tool to create a 5-mile radius around your home. Then select all the parks that fall within the buffer zone to see what's nearby. (See Chapter 9 for details on finding features with buffers.)

5. **To create a simple map,** add labels to show park names and adjust colors and symbols for clarity.

   Export your map as a PDF or image to share with friends or community members. (See Chapter 17 for cartographic design tips.)

3
Retrieving, Counting, and Characterizing Geography

IN THIS PART . . .

Uncover the hidden details in raster data and learn how to analyze and visualize patterns across the landscape.

Explore how to calculate basic summary statistics and interpret patterns in geographic data.

Discover how to find geographic features based on their attributes, spatial relationships, and patterns.

> **IN THIS CHAPTER**
> » Discovering the details hidden in your raster data
> » Searching with GIS tools
> » Making your results make sense

Chapter 8
Exploring the World through Raster Data

Raster data lets you view the world as a patchwork of pixels, each one brimming with unique details about the landscape below. From broad forests and winding rivers to subtle elevation changes or tiny urban patches, rasters offer a way to analyze large areas while digging into the details hidden in every pixel. Whether you're identifying peaks, tracing roads, or calculating areas at risk for climate change, this chapter walks you through how to find and interpret the valuable insights within your data.

With a few handy techniques, you discover how to search for features, calculate stats, and even apply advanced analysis to uncover patterns and trends. Raster data not only helps you see the landscape but also tell its story. So dive in and start exploring how to make the most of this pixel-powered perspective!

Identifying and Locating Features in Raster Data

When analyzing raster data in GIS, you're working with a pixel-based view of the world, in which each pixel holds valuable information about the landscape below. Imagine each pixel as a tiny data point representing features that may span miles, like a dense forest or flowing river, or more specific locations such as a park or landmark. Raster data allows you to explore huge areas, but finding and identifying specific features requires you to take a deeper look at what's hidden in each pixel. (To find out more about rasters, check out Chapter 5.)

In this section, I guide you through practical techniques for locating and analyzing different types of features in raster data — whether you're searching for a point of interest, tracing linear features like roads and rivers, or analyzing large regions with unique characteristics. Using these techniques, you discover methods for pinpointing exact locations within a pixel grid (raster), tracing lines to represent roads and rivers, and examining broader areas to see how vegetation, urban development, or other types of coverage change across the landscape. Each type of feature has its own quirks, often requiring a unique approach. Understanding the quirks and how to work with them will help you unlock the full potential of your raster data, enabling insights that can inform decisions, guide planning, and support a variety of geospatial analyses.

Locating areas of interest

One of the cool things about raster data is that it's like a treasure map, with each pixel holding clues about the landscape. Your "treasure" might be finding the highest peaks in a mountain range or pinpointing valleys and lowlands for a flood-risk study. In cases like these, a *digital elevation model (DEM)* — a special raster layer that provides elevation — is your go-to raster. For example, Figure 8-1 shows a DEM of a small mountain range, and the hatched area in the image on the right depicts areas above 7,000 feet.

Imagine that you're conducting a watershed study. Your first task is to identify regions above 7,000 feet so that you can focus on snowpack areas that are important for downstream water supplies. You can easily find these areas using GIS tools and a DEM, as shown in Figure 8-1.

To locate areas of interest in a raster, you follow these steps:

1. **Choose the right raster layer.**

   Start by figuring out which layer has the data you need. For elevation analysis, that's your DEM layer. If you were studying vegetation, you would probably use a land cover layer, and for urban heat island studies, a temperature raster layer would probably be best.

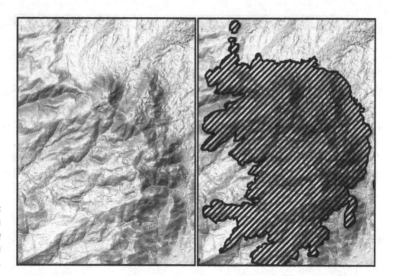

FIGURE 8-1:
A digital elevation model with the higher elevation area highlighted.

2. **Define your target values.**

   Decide what values you want to look for in the data. For elevation, you might set a threshold of 7,000 feet, as I did in Figure 8-1. You use GIS to flag any pixels over that value as your high points.

3. **Run a spatial query.**

   Use a spatial query tool in your GIS software to search for pixels that match your values or fall within your determined range of values. In QGIS, you can use the Raster Calculator, whereas ArcGIS Pro offers two options for performing this step: the Raster Calculator and the conditional evaluation (or Con) tool. The Raster Calculator allows you to isolate pixels that match your criteria. In Figure 8-2, I used QGIS to flag all pixels with an elevation value greater than or equal to 7,000 feet.

4. **Check and save your results.**

   Take a look at the search results. Make sure the areas identified are accurate and make sense for your project. You can save these areas as a new layer if you want to keep them for further analysis.

TIP
Running spatial queries doesn't just help you find areas of interest but can also help you visualize patterns in your data. By identifying peaks, valleys, or vegetation types, you reveal the hidden treasure — the "story" of the landscape and how these features are arranged across the area. Avast, me matey; treasure found!

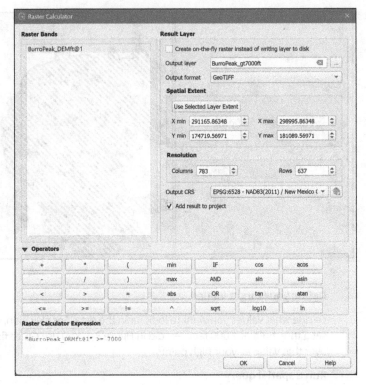

FIGURE 8-2: Using the Raster Calculator in QGIS.

Analyzing linear feature

Linear features, such as rivers or roads, create natural connections across a landscape, linking areas together like paths on a map. In many cases, these features are first identified using raster data because satellite imagery and elevation models — which are common sources for detecting natural and human-made paths — are often stored as rasters. In raster format, linear features appear as chains of pixels forming a continuous path, but this representation is not always ideal for analysis.

Because raster data are stored as pixel grids, they lack the precision, scalability, and flexibility of vector data. Converting raster features to vector format allows for more detailed spatial analysis, such as measuring distances along roads, modeling river networks for hydrological studies, or finding the most efficient travel routes. This conversion also improves visualization, making maps cleaner and easier to interpret.

To identify linear features in a raster, and then convert and analyze them in vector format, follow these steps:

1. **Locate the lines.**

   Use edge detection tools in your GIS to find the boundaries for roads, rivers, or other pathways. These tools detect sudden changes in pixel values, helping you locate edges that reveal linear features. In ArcGIS Pro, use raster functions such as the Convolution tool for edge detection; in QGIS, start with the `slope`, `aspect`, and `hillshade` functions in the Terrain Analysis toolbox.

2. **Extract the line features.**

   Use line-extraction tools to convert pixel-based paths into vector lines. Doing so creates a continuous, smooth line representing your feature, whether it's a winding river or cross-country highway, and provides a cleaner, easier-to-read version that's ready for analysis. For large datasets, you can automate the process, which I cover in Chapter 18.

3. **Classify and analyze.**

   With your lines defined as vector features, you're ready to dive into the analysis. Now you can find the shortest route, trace paths across regions, or assess how these features interact with the surrounding landscape.

TIP

Edge detection and line-extraction tools are great for finding and simplifying linear features like rivers, streams, and roads. Converting these features into smooth vector data makes it easier to perform analyses like planning evacuation routes or assessing flood-prone areas based on proximity to rivers and streams.

Exploring areas and distributions

Raster datasets are especially useful for exploring large areas and spotting spatial patterns. Whether you're mapping forest boundaries, wetlands, or urban sprawl, raster data help you see how different areas are arranged across the landscape. Studying these areas gives you insights into trends, changes over time, and ways to plan for future needs.

To find areas and distributions in raster data, follow these steps:

1. **Choose your focus by selecting the layer that represents the feature you're interested in, such as land cover or vegetation.**
2. **Classify by using clustering techniques.**

   Use classification tools in your GIS to group similar pixels, creating distinct classes like "forest," "water," or "urban area." You can use supervised classification (in which you guide the process) or unsupervised classification (the software groups pixels based on patterns). Each class gives you a clearer picture of where specific types of land cover or land use are distributed, as shown in Figure 8-3. (You can't see the colors if you're reading the print book, but go to https://www.mrlc.gov/viewer/ for examples).

3. **Track changes and relationships.**

    After you classify the pixels, use the resulting data to start tracking changes over time or exploring relationships with other features. For example, you might notice vegetation zones expanding or urban areas encroaching on previously undeveloped land.

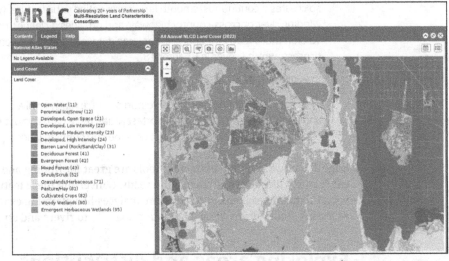

FIGURE 8-3: Land-cover classifications in the Lower Klamath National Wildlife Refuge.

From the All Annual NLCD Land Cover (2023) - Klamath Basin National Wildlife Refuge Area (41.82249 lat, -121.80180 lon) & Map from MRLC / Public Domain

REMEMBER

Watching distributions change over time reveals long-term trends, making it easier to adjust plans for conservation, land management, or urban development.

Performing Searches in Raster Data

Raster data analysis used to come with a variety of limitations, both in capabilities and user accessibility. Limited to simple spatial queries and the most basic spatial analysis functions, raster data was cumbersome and frustrating to work with. But luckily for you (and me!), working with raster data in GIS has evolved. Modern platforms help you put those frustrations behind you, making it easy and efficient to find, filter, and analyze raster data.

Searching in simple rasters

In a raster, searching is a lot like filtering a photo to highlight specific details. Each pixel holds a value, like a color or number, and you can use simple tools to bring out the features you're interested in. With a few basic GIS techniques, you can find, highlight, and isolate parts of your data. Here are three essential tools for performing quick searches in rasters:

>> **Reclassification:** This technique groups pixels with similar values into general categories. For example, if you're working with elevation data, you might group pixels into categories like "Low," "Medium," and "High" instead of trying to interpret hundreds of different values. Reclassification is especially handy if you need to simplify your data for visualization or further analysis.

>> **Thresholding:** Thresholding lets you filter pixels based on specific values by setting a cutoff point. For instance, when working with a temperature raster, you can set a threshold to show only areas above 90°F (32°C). This technique is great when you're looking for extremes or specific ranges in your data.

>> **Attribute-based queries:** With attribute-based queries, you search for pixels that match specific values, much like searching for terms in a spreadsheet or database. In this case, you're searching for attribute information attached to each pixel: temperature, elevation, or land cover type. For example, on a land cover map, you can isolate and calculate the area of all the wetlands by querying for pixels with the value for wetlands.

These basic raster search tools make it easy to highlight and isolate features in your data, letting you focus on areas of interest. Whether you're working with natural landscapes, urban areas, or environmental data, these techniques allow you to see patterns, trends, and features clearly, even in simple raster formats.

Searching DBMS and cloud-supported rasters

With data constantly increasing, cloud storage and database management systems (DBMS) make it possible to work with larger datasets efficiently. Platforms like PostgreSQL and PostGIS allow you to store, manage, and query vast amounts of raster data, giving you flexibility to manage data across different locations. Need a massive dataset of global temperature changes? PostGIS or a cloud-based system like Google Earth Engine can help you store and process data without maxing out your computer's memory and hard drive space.

REMEMBER

With DBMS-supported rasters, you get powerful search options that go beyond basic reclassification and thresholding. You can search based on complex attributes, calculate area coverage over time, or filter for specific criteria across multiple datasets.

Using machine learning and AI for searches

Machine learning (ML) and artificial intelligence (AI) tools have unlocked new ways to search for and detect patterns in raster data. Techniques like convolutional neural networks (CNNs) allow for automated feature detection and pattern recognition, which is especially useful for identifying specific land cover types, vegetation health, or even urban infrastructure.

Imagine needing to identify buildings in satellite images for urban planning. Instead of manually tagging each building, ML models can detect building-like structures automatically. This capability not only saves time but also offers a consistent and fairly accurate way to identify features across vast landscapes. With machine learning, raster searches have become faster and more efficient, helping you find valuable insights that would be impossible to achieve manually.

WARNING

Although ML and AI can significantly speed up your work, it's important to "trust but verify" the results. Automated tools aren't perfect and can misinterpret features, such as by tagging dry washes as roads or mistaking shadows for bodies of water. Always double-check critical results to ensure that they're accurate and safe to use, especially in applications like routing or hazard assessments. A little extra caution can go a long way in avoiding potentially dangerous or costly mistakes.

Calculating Statistics and Summarizing Data

Finding the features you're interested in is one thing, but the fun truly begins when you get to dig into the numbers. Summarizing your data with statistics isn't just about crunching numbers; it also helps to reveal patterns and stories hidden within, like spotting climate shifts over decades or understanding where wildlife habitats overlap with developed urban areas. In this section, I guide you through the essentials plus give you a look at some advanced techniques. (If you're ready to get nerdy, jump over to Part 4, where I show you a variety of methods for analyzing geographic patterns.)

Getting simple statistics

GIS is powerful not only for finding and visualizing features but also for its capability to quickly generate basic statistics, revealing patterns and characteristics of your data. For instance, if you're a city planner, knowing the average elevation across the city can help inform flood management strategies.

Simple statistics, like minimum, maximum, mean, median, and mode, give you a quick look at trends across your dataset, and standard deviation can help you understand variations in your data. These basic statistics are essential for interpreting everything from temperature changes to vegetation density. Here's a brief look at each type of statistic:

» **Minimum and maximum:** These values give you the range of your data. If you're working with elevation data, you can get an idea of the landscape's overall variation by knowing the lowest valley (minimum) and highest peak (maximum). Understanding these extremes can be crucial for risk assessment and environmental planning.

» **Mean (average):** The mean gives you the average value across all pixels in raster. If you're working with temperature data across a region, the mean can give you a sense of the area's general temperature. The mean is particularly useful for comparing regions or time periods because you can see how one area's average compares to another's.

» **Median:** When you arrange your pixel values in order from lowest to highest, the median is the one that falls exactly in the middle. In contrast to the mean, the median isn't affected by extreme values at one end or the other, making it useful for datasets with outliers, like a generally flat area with just a few very high elevation points.

» **Mode:** The mode is the value in your dataset that occurs most often. This statistic is helpful when working with categorical data such as land cover types. Knowing the mode in a land-cover dataset might tell you which type of vegetation is most common across an area, whether it's forest, grassland, or desert scrub.

» **Standard deviation:** This statistic shows how much variation there is from the mean. If you're working with temperature data, a high standard deviation indicates large swings in temperature across the area, whereas a low deviation means that temperatures are more uniform. This information can help you identify regions with consistent conditions versus those with high variability.

These simple statistics give you a way to summarize raster data in a meaningful, digestible way. Many GIS programs will compute these values for you on the fly, saving time and providing insights at your fingertips for use in planning, analysis, and further studies.

TIP

Try experimenting with each of these statistics on sample raster data to see how they shape your interpretation. Summarizing data in this way doesn't just answer questions about individual features but also helps you grasp the larger patterns and variations across your data. Check out NASA's Earthdata online portal for global temperature and climate data.

Exploring advanced statistical techniques

If you want to dig deeper into your raster data, advanced statistical techniques like map algebra and its component, zonal statistics, help you go beyond individual data layers to see how they interact.

Map algebra is, as the name implies, a way to perform math with raster layers by adding, subtracting, or multiplying values in each cell. Map algebra lets you calculate across layers, like combining temperature and elevation to pinpoint frost-prone zones. For example, to assess climate impact, use zonal statistics to calculate the average temperature in different vegetation zones. Both map algebra and zonal statistics are powerful tools for raster analysis. Check out Chapter 16 for a deeper dive into map algebra and its component functions.

Advanced statistics add a new layer of analysis, allowing you to connect data across different themes, examine trends over time, and understand the relationships between features.

Visualizing and interpreting raster data

Generating statistics is just one part of analyzing your data. The other part is visualizing it. Visualization methods help you interpret and communicate your data. You can display raster data in a variety of ways. Hillshades, gradients, and classifications are popular options for displaying raster data because they show patterns in a way that's easy to understand at a glance. For instance, refer to Figure 8-3, earlier in the chapter, to see land cover classifications. Visualizing land cover data on a map lets you quickly spot agricultural areas within and around wetlands, making patterns that might be missed in raw data easily visible to both you and your audience.

Visualization is a powerful way to communicate patterns in your data, showing where values cluster or spread. Visualization tools make complex results accessible, helping others quickly grasp the big picture of your analysis.

Reporting and sharing results

After all your hard work of collecting, tabulating, and preparing statistics, it's time to share your findings! Whether you're preparing a printed report or creating an interactive dashboard, clear visuals and summaries ensure that your analysis is accessible to all. With GIS, you have options like color-coded maps, interactive dashboards, and web maps, making it easy for anyone to view your data — no special software required.

Imagine that you've calculated areas at risk for flooding. Instead of presenting a bunch of text and detailed tables of raw data, you bring your data to life by creating a report with color-coded maps, clear charts, and summary tables. These visual aids make it easy for your audience to spot high-risk zones and understand your recommendations.

When you tailor your reports to your audience — whether by creating a printed report for planners, an interactive dashboard for local officials, or an online interactive map-based app for the public — you're helping your audience not only understand but also act on what you've found. A great report doesn't just look good; it makes your data meaningful.

Make sure that your visuals do the talking! Clear, organized reports with great maps and easy-to-read charts help your audience connect with your findings to get the full picture — literally! For more tips on creating effective output, check out Chapter 17.

> **IN THIS CHAPTER**
> » Understanding vector data structure
> » Harnessing spatial relationships for precise searches
> » Analyzing and visualizing search results
> » Ensuring that your data is accurate and reliable

Chapter 9
Finding Features in Vector Data

Vector data is the precision instrument of the GIS world, offering a flexible, detailed view that lets you trace boundaries, measure distances, and explore connections across any landscape. Unlike raster data, which provides a broad, pixel-based picture, vector data zooms in on exact locations and precise shapes. Think of the difference between raster and vector as switching from a bird's-eye view to a detailed street map in which every road, building, and boundary has a specific place.

In this chapter, I show you how vector data transforms GIS analysis, allowing you to find the exact features you're interested in, answer targeted questions, and uncover relationships that may be hidden in your data. Whether you're searching for the nearest park, tracing complex road networks, or analyzing property boundaries, vector data offers the tools to get detailed, reliable answers. In short, this chapter tells you what makes vector data so powerful and how to harness it effectively.

Getting Explicit with Vector Data

When working in GIS, the type of data you choose makes a big difference in how precisely you can represent the real world. In contrast to raster data, which consists of a collection of pixels and is great for capturing broad landscapes, vector data lets you get specific, enabling you to pinpoint exact locations and draw precise lines and shapes. Because vector data lets you work with geographic features in a more realistic way, it's often your go-to choice for detailed mapping projects.

REMEMBER

Think of vector data as the drawing tools of GIS. Instead of using pixels, it relies on three basic shapes — points, lines, and polygons — to represent real-world features with accuracy and clarity. Here's what each basic vector type is best suited for:

» **Points:** Points are perfect for indicating exact locations, like a bus stop, well, or trailhead. Each point represents a specific spot, so you can easily plot places of interest or identify locations in a clear, organized way.

» **Lines:** With lines, you can connect points to map linear features such as roads, rivers, or power lines. Lines are essential for showing paths and networks across the landscape. For example, if you're mapping a road network, lines let you trace exact routes, making it easy to analyze connectivity or calculate shortest paths. Along these routes, points can mark key locations such as starting points, waypoints, and end points, providing critical details for navigation and route analysis.

» **Polygons:** You use polygons to define areas with boundaries. These areas can represent anything with a distinct shape and size, from building footprints to lakes to land parcels. If you're an urban planner and need to understand how much land is zoned for residential use, polygons provide exact boundaries, making calculations and area-based analysis easy.

Say you're an urban planner tasked with planning a new bike path through the city. With vector data, you can draw precise routes using lines, measure distances accurately, and avoid obstacles represented as polygons. Or imagine you're assessing bird habitats. Points let you mark specific nesting sites, ensuring that exact locations are captured for your analysis. These examples highlight how vector data's precision and flexibility allow you to work with real-world features in a way that's both practical and powerful. Whether you're designing bike paths, protecting wildlife, or analyzing urban spaces, vector data gives you the tools to map and measure with confidence, making it the backbone of many GIS projects.

Seeing How Data Structure Affects Retrieval

Vector data is more than just points, lines, and polygons; it's a data structure designed to make finding and analyzing features efficient and powerful. The way your vector data are organized influences how quickly and accurately you can get the information you need from it. Under the hood, vectors have aspects like topology and spatial indexing that help your GIS easily perform complex tasks.

Building relationships

In GIS, vector data follow a set of rules and relationships that define how points, lines, and polygons interact. Features don't just float around on a map; they are structured in a way that helps GIS understand spatial relationships, such as which features are connected, adjacent, or contained within others. This topological framework (described in Chapter 5) enables GIS to efficiently perform tasks like calculating routes, finding nearby features, and analyzing spatial patterns.

Topology is a fundamental concept of GIS, providing a structured way to define spatial relationships between features. Understanding its three main components — connectivity, adjacency, and containment — can help you make better sense of how spatial data behave.

Topology provides the foundation that GIS uses to analyze spatial relationships. Whether you're determining which streets are connected, identifying adjacent land parcels, or finding parks within city limits, a strong topological structure ensures that GIS produces accurate results.

Optimizing your data for speedy searches

When working with large datasets, even well-structured topologies benefit from optimization. One way in which GIS improves efficiency is through *spatial indexing*, a data structure that organizes spatial features to speed up searching. Instead of combing through every point, line, or polygon in a dataset, spatial indexing creates a "shortcut" that helps GIS quickly locate relevant features based on their geographic location.

Imagine that you're working with a statewide map of hiking trails. Rather than manually (and painfully!) reviewing every trail to find just the ones in your area, spatial indexing allows GIS to pinpoint relevant features instantly. Using spatial indexing is like hitting fast forward so that the software can jump to the right data

CHAPTER 9 **Finding Features in Vector Data** 135

without wading through unnecessary details. This efficiency is especially crucial when dealing with massive datasets in enterprise or cloud environments, ensuring fast, accurate searches without unnecessary delays.

REMEMBER

Whether you're looking for coffee shops, trails, or the layout of an entire city, vector data's structure and topology give GIS the smarts it needs to pull up the right results quickly and accurately. When your data is well structured and indexed, you'll get answers faster and with fewer hiccups. Topology and spatial indexing are like your GIS toolkit's pit crew, ensuring that everything runs smoothly so that you can focus on driving.

Choosing the Right Data Source

When working with vector data, you often need to search for specific features or patterns such as finding parks within a city, locating utility lines, or identifying nesting areas within wetlands. Vector data's capability to represent precise locations, shapes, and connections makes it perfect for these kinds of detailed analysis. But to get meaningful results, you need to start with the right data.

Choosing the right data source is like picking the right tool for the job. Each data source brings its own strengths, and selecting the wrong one can lead to frustrating roadblocks or inaccurate results. By considering factors like data type, scale, accuracy, and precision, you set yourself up for success, ensuring that your searches yield meaningful insights and align with your project goals.

Targeting the right data

Your project goals drive which vector data is best for you to search. Different projects call for different types of data, and targeting the right data keeps your searches on track and your results meaningful.

For example, if you're analyzing flood risks, hydrography data with medium-scale detail is ideal for identifying rivers and streams. But if you're planning wildfire response strategies, high-accuracy road-network data is essential to route emergency vehicles quickly and safely.

Keep the following four key factors in mind when choosing your data sources:

» **Data Type:** Identify what kind of information — points, lines, or polygons — you're working with. Each data type has strengths for different kinds of analysis, so you want to base your choice on what your project needs most:

- Points depict specific locations, like fire hydrants and wells.
- Lines connect features, like roads and rivers.
- Polygons represent areas, like parks and lakes.

» **Scale:** Determine how "zoomed in" you need to be. Scale is about the level of detail in your dataset. For example, state-wide data (small scale, less detail) won't show the individual buildings and parcels that you'd find in neighborhood-level data (large scale, more detail). For more details about scale, check out Chapter 2.

» **Accuracy:** Estimate how close your data's locations and measurements are to the real world. For example, a trailhead mapped at an accuracy of ten feet might be close enough for a hiking app but too vague for land surveying. Choose data with accuracy that matches your project's needs.

» **Precision:** Be aware of how much detail you need. Precision refers to the level of detail in the data, such as the number of decimal places. High precision gives consistent, specific values but doesn't guarantee accuracy. You can have coordinates with six decimal places that are consistently off by 100 feet — which is highly precise but not very accurate.

REMEMBER

It's easy to confuse accuracy and precision. Think of it this way: Accuracy is telling the truth; precision is telling the same story the same way over and over. Just because your data is precise doesn't mean that it's accurate, and just because it's accurate doesn't mean that it's precise. In GIS, high-precision coordinates or measurements that aren't accurate will still miss the mark if they're off from the real location. Figure 9-1 shows the differences between accuracy and precision.

FIGURE 9-1: Accuracy versus precision.

High accuracy High precision   Low accuracy High precision   High accuracy Low precision   Low accuracy Low precision

ChemistryGod/Adobe Stock Photos

Whether you're monitoring wildlife migration corridors, planning public infrastructure, or managing conservation efforts, selecting the right data can make all the difference in achieving accurate, actionable results. Table 9-1 highlights how data type, scale, accuracy, and precision come into play for a variety of use cases.

TABLE 9-1   Sample Data Use Cases

Use Case	Ideal Data Type	Rationale for This Data Type
Emergency response planning	Points (emergency services); small scale; high accuracy and moderate precision	Pinpointing locations with high accuracy ensures that hospitals and fire stations are correctly mapped for emergency response; moderate precision is sufficient because responders need general locations rather than exact, centimeter-level details.
Transportation analysis	Lines (roads and streets); medium scale; moderate accuracy and precision	Road networks show connectivity essential for routing and infrastructure planning. High accuracy and precision are crucial for building roads, but moderate accuracy and precision are sufficient for planning and analysis, for which general alignment and connectivity matter more than exact placement.
Land development studies	Polygons (property boundaries and parcels); large scale; high accuracy and precision	High accuracy and precision ensure that property boundaries and parcels are correctly mapped for zoning, legal boundaries, and property development planning, for which exact placement is critical for decision-making.
Wildlife habitat mapping	Polygons (protected areas); medium scale; high accuracy and moderate precision	Wildlife habitat mapping relies on accurate spatial data to define conservation zones and assess habitat overlaps. Although high accuracy ensures that habitats are correctly located, moderate precision is often sufficient because general boundaries and ecological patterns matter more than exact placement.
Water resource management	Lines (hydrography datasets); medium to large scale; high accuracy and precision	Hydrography datasets provide detailed representations of rivers, streams, and water bodies. High accuracy and precision are essential for flood-risk assessments, watershed management, and water resource planning.

Predicting the outcome

Wouldn't it be great to have a crystal ball to look into and know exactly how your final map should look? Choosing the right data and defining your searches would be a snap! Although crystal balls may be hard to come by these days, you have the next best thing: A clear vision of your end goal. Thinking ahead about the final outcome acts as your crystal ball, helping you pinpoint the right data and tools to bring your vision to life.

TIP

Try this "crystal ball" approach to keep yourself on track:

1. **Set your vision.**

    Define your goals in geographic terms. For example, do you want a close-up of neighborhood park access or a big-picture view of city green space?

2. **Choose your tools.**

   Decide which spatial features (points, lines, or polygons) will bring that vision to life.

3. **Focus your search.**

   Use your data's scale, type, and accuracy as guides to refine your results.

4. **Analyze with purpose.**

   Compare your results to your original vision to be sure they match.

By "predicting the outcome" and working backward, you create a clear roadmap for each step of your project. No crystal ball, spells, or potions required — just a little planning and some GIS magic!

Locating Specific Features with SQL

Finding specific features like cities, parks, or lakes in your data is a breeze with Structured Query Language (or SQL, commonly pronounced "sequel" although it's second to none). SQL is your behind-the-scenes query assistant, letting you ask precise questions of your data to get exactly the answers you need. Although with traditional SQL, you had to write commands using terms like SELECT and WHERE, modern GIS tools simplify the process through intuitive interfaces, so you don't need to memorize code to create powerful queries.

The modern tools translate your searches into SQL queries automatically, giving you the best of both worlds: a properly structured SQL command and ease of use. In this section, I show you how to use these user-friendly tools to search your data effectively. For an even deeper dive into searching, see Chapter 10.

Using SQL queries based on attributes

One of the best parts of modern GIS software is that you don't need a programming degree to write effective SQL queries! Instead of typing SQL commands, you simply build queries step by step using a graphical interface. Follow these steps to build a query:

1. **Select the data layer with the features you're interested in.**

   For example, if you're looking for high-population cities, choose the cities layer.

2. **Open and browse the attribute table to understand the field names and values available for building your query.**

   The attribute table is the "cheat sheet" for your query. Figure 9-2 shows an attribute table for a point layer of cities in Africa.

3. **Open the attribute selection tool in your GIS software.**

   Depending on the GIS software you're using, this tool might be called "select by attributes" (ArcGIS) or "select features by value" (QGIS). Figure 9-3 shows the selection tools in ArcGIS Pro and QGIS.

4. **Build your SQL query by using a simple SQL statement.**

   For example, to find high-population cities, you might use Population > 1000000 to find cities with more than a million people.

5. **Run the query by clicking OK**

   Now you get to see the magic happen. Selected features will be highlighted on your map and in the attribute table, making them easy to view and analyze.

6. **Use the selected features to visualize trends, update attributes, or save them for future analysis.**

   More details on saving them later in this chapter.

   Modern GIS software does the heavy lifting behind the scenes, ensuring that your queries are powerful and easy to use.

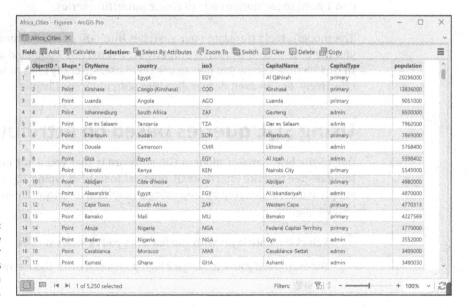

FIGURE 9-2: An attribute table for a feature layer of African cities (shown in ArcGIS Pro).

140　PART 3　Retrieving, Counting, and Characterizing Geography

FIGURE 9-3: Attribute selection tools in ArcGIS Pro (top) and QGIS (bottom).

Exploring advanced SQL functions

SQL is more than just basic searches. As the following list details, with advanced functions like joins, spatial joins, aggregations, and advanced filtering you can link datasets, analyze relationships, and generate summary statistics for deeper insights:

» **Join:** Suppose that you have two tables; one is an attribute table for city location points and the other is a stand-alone table containing population data for those cities. A *join* links those two tables, allowing you to run queries on both of them as if they were one. For example, with population data joined to the city attribute table, you can identify the most populous cities in a region, calculate average population density, or even symbolize the city points based on population size (larger dots for higher population). Doing so makes it easier to uncover patterns and identify hot spots that wouldn't be possible if the data remained separate.

» **Spatial join:** In contrast to a table join, a spatial join links datasets based on geographic relationships. For example, you can join parcels to zoning layers to determine which parcels fall within specific zoning types, or you can connect school points to attendance boundaries to assign each school to its respective zone.

» **Aggregation:** Aggregate functions like SUM, AVG, and COUNT allow you to calculate totals, averages, or counts across a dataset. For example, you use aggregations to count the number of parks in a city, calculate the total road length in a county, or determine the average rainfall across counties.

» **Advanced filtering:** You can simplify your analysis by displaying specific subsets of your data. You do so by using definition queries (ArcGIS Pro) or expression filters (QGIS) to create virtual filters without altering the original dataset. For example, for a school-related dataset, you might apply a filter like School_Type = 'Elementary' to focus on just elementary schools while keeping the other data intact.

REMEMBER

Advanced SQL functions are like your GIS secret weapon — powerful but easy to manage and use with modern tools. Don't be afraid to experiment with joins, filters, and aggregations to uncover patterns and insights that might otherwise stay hidden in your data!

Making the most of your search results

Your search results serve as more than just a quick answer to a query; they also provide the foundation for deeper analysis. What you do with your results can take your analysis to the next level. Whether you want to explore patterns, update attributes, or create a new dataset, try these tips for getting the most mileage out of your results:

» **Spot patterns at a glance.** Highlight clusters or trends in your data, like densely populated areas or regions with abundant green space.

» **Update attributes in bulk.** Save time by editing multiple records simultaneously. For instance, if you've identified schools with outdated enrollment data, you can update those fields all in one shot.

» **Save your results as a new layer in your dataset.** Turn your selection into a permanent dataset for further analysis. Here's how to do it in ArcGIS and QGIS:

- **In ArcGIS Pro:** Right-click the layer in the Contents pane, choose Data ⇨ Export Features, and ensure that the Use Selected Records option is toggled on before saving your layer.

- **In QGIS:** Right-click the layer in the Layers panel, choose Export ⇨ Save Selected Features As, and choose your desired format and location. Figure 9-4 shows the dialog box for exporting selected features with QGIS.

Give your new layer a clear, descriptive name, like `CitiesOver1M` or `FloodRiskZones`, so that you can easily find it later.

» **Export for sharing and collaboration.** If you need to share your selected features with others or use them in different software, exporting to common file formats ensures compatibility with other GIS tools or workflows. Common shareable formats include the following:

- **Shapefile (`.shp`, `.shx`, `.dbf`):** Widely supported but has limitations like a ten-character attribute field name cap. Also, be sure to include all files associated with the `.shp` (despite the name, shapefiles are made up of multiple files; see Chapter 5 for more about shapefiles).

- **GeoJSON (`.geojson`):** Great for web-based applications and lightweight GIS analysis.

- **GeoPackage (`.gpkg`):** A modern all-in-one format that stores multiple layers, attributes, and even raster data in a single file. It's highly efficient, open standard (freely available rules that ensure compatibility across different software), and works seamlessly with ArcGIS Pro, QGIS, and other platforms.

- **File Geodatabase (`.gdb`):** Best for ArcGIS users needing a complete package of vector, raster, and tabular data.

- **KML (`.kml`):** Ideal for sharing with non-GIS users through Google Earth or Google Maps, with visualization-ready features.

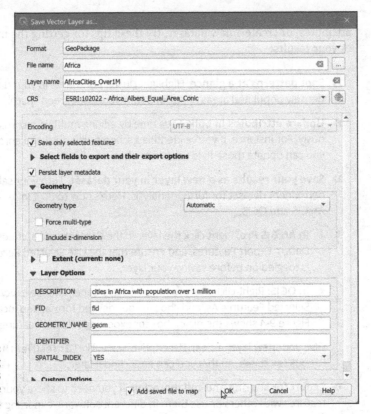

FIGURE 9-4: Exporting selected features to a GeoPackage with QGIS.

REMEMBER

When exporting data for sharing, tailor the format to your audience. A GIS-savvy colleague might prefer a GeoPackage or Geodatabase, whereas a city planner might find a KML or interactive web map more useful.

TIP

GeoPackage is a great go-to format for sharing because it avoids the clutter of multiple files (like shapefiles) and supports advanced GIS functionality.

Making the most of your search results not only saves time but also enhances efficiency and collaboration. Whether you're updating, analyzing, or sharing, these techniques ensure that your searches are more than just a one-time task; they become building blocks for deeper insights and streamlined projects.

Searching Vector Data with Geography

Vector data helps you understand *spatial relationships*, which refers to how features connect, overlap, or interact with one another. Working with vector data is what truly brings GIS to life, allowing you to uncover insights like which parks are near schools, which rivers cross county boundaries, or which neighborhoods fall within a floodplain.

Using spatial operators

Spatial operators are the power tools for analyzing spatial relationships. Spatial operators like within, intersect, or near are built into most GIS software, making spatial queries straightforward and user friendly. These power tools help you answer questions like the following:

» Which parcels are *within* the city limits?

» Which roads *intersect* this planned development area?

» Which schools are *near* the proposed housing project?

Imagine that you're a city planner trying to figure out which schools are within walking distance of a park. Spatial operators like within do the heavy lifting for you, saving you hours of manual counting. Figure 9-5 shows parks selected within the city of Frisco, Texas. You can accomplish the same thing using these steps in ArcGIS Pro:

1. **Start with a new map project and load the city boundary layer and the parks layer.**

   The layers will display in the map view and be listed in the table of contents.

2. **On the Map tab, choose Select by Location in the Selection group.**

   The Select by Location dialog box opens, as shown in Figure 9-5.

3. **In the Select by Location dialog box, choose Parks as the Input Features; this is the layer containing the features that you want to select.**

   The layer you're selecting features from, Parks (in this example), should appear in the Input Features selection box.

4. **For Relationship, select Within from the drop-down menu.**

   This is where you indicate the spatial operation you're performing, which in this case is finding parks within the city.

5. **Choose the city layer you added in Step 1 as the Selecting Features.**

   In this example, the city layer is the polygon boundary layer for Frisco, Texas.

6. **Click Apply or OK to run the query.**

   The park points that fall within the city polygon boundary should now be highlighted on your map.

7. **Review the selected features and save the selected features as a new layer if needed.**

REMEMBER

Operators like `within`, `intersect`, and `near` make it easy to query data based on how features relate to each other in space.

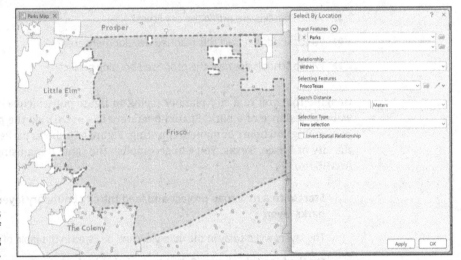

FIGURE 9-5: Selecting parks within the city of Frisco using ArcGIS Pro.

Finding features based on distance or buffers

A *buffer* is another helpful tool to use for spatial searches. A buffer is based on distance and is used for defining areas of interest, such as for identifying all homes within a two-mile radius of a school or finding all coffee shops within a ten-minute walk of a location. Here are some basic steps for creating a buffer around a school:

1. **Start a new project in your GIS software and add the schools layer.**
2. **Select the school you want to buffer using the Selection tool.**

3. **Open the Buffer tool:**
   - **In ArcGIS Pro:** Search for the Buffer tool in the Geoprocessing pane.
   - **In QGIS:** Right-click the layer, go to Geoprocessing Tools, and select Buffer.
4. **Enter the buffer distance (for example, 2 for two miles).**
5. **Choose whether the buffer should overlap other buffers.**

   Use the Overlap option if you're creating buffers around multiple features or schools, as in this example.

6. **Click Apply or Run (or whatever works in your software) to run the tool to create the buffer as a new polygon layer.**

   You can now use your new buffer layer to find features within the buffer, like houses, parks, or bus stops.

Buffers are especially useful for

» Identifying flood-prone areas near rivers

» Finding houses within a certain distance of a proposed highway

» Highlighting wildlife zones within proximity to industrial areas

Buffers let you define areas of interest to focus your analysis on, where it matters most. For example, an urban planner can buffer bus stops to see which neighborhoods are underserved by public transportation.

TIP

Buffers aren't just circles! Although a uniform buffer around a school point creates a perfect circle, a variable buffer along a winding river adapts to its curves, ensuring more precise results. Don't hesitate to play around with buffer settings in your GIS software. For example, try overlapping buffers or adjusting distances to see how your results change.

Counting, Tabulation, and Summary Statistics

After you've identified the features that you need for your analysis, as described in the previous sections of this chapter, the next step is to dig into the details of the data and any selected features. Counting, summarizing, and visualizing data are essential steps to understanding the story that lies within your data.

Performing basic statistical analysis

With its associated attribute table, vector data makes it easy to count features and calculate basic statistics. Imagine being tasked with managing the emergency responses for a city. Counting fire stations within districts can quickly highlight areas that are lacking resources. You can use basic statistics on your GIS data in a number of ways, such as to

» Sum the total length of roads in a transportation layer

» Determine the average number of households per neighborhood

» Calculate the area of land parcels for tax assessments

GIS software, which often comes with built-in summary statistics tools, makes these tasks straightforward. For instance, field calculations can quickly provide totals, averages, or even custom metrics for selected features. Refer to the section in Chapter 8 about calculating statistics and summarizing data for details on the different statistics you can pull from your data.

Visualizing and summarizing data

Visualizing your results is where your data, and the story it tells, comes to life. Dashboards and charts can help communicate trends and patterns clearly. For example:

» Bar charts help city planners determine which neighborhoods have more parks and which ones need them.

» Pie charts help economic developers quickly see the proportion of land use types in a city, providing insights for targeted growth.

» Heat maps provide businesses with insights into where clusters of service requests occur, helping them allocate resources efficiently.

Modern GIS platforms let you combine these visualizations with interactive maps, creating a compelling way to share your findings with decision-makers, stakeholders, and the public. Visualizing results not only makes your data easier to interpret but also helps stakeholders quickly grasp critical trends and patterns, empowering better decisions.

TIP

Look into tools like ArcGIS Dashboards, Experience Builder, Tableau, Power BI, or the variety of visualization libraries for Python or R. These tools help you create visual summaries of your analysis that are both informative and visually appealing. Also check out Chapter 17, where I cover a variety of ways to generate output with GIS.

Validating and Verifying the Results

No matter how powerful your tools are, taking a moment to double-check your results is always worthwhile. Even small errors can lead to big misunderstandings. Validation ensures that your results are accurate, complete, and suitable for your goals. Start by asking yourself these critical questions:

» **Did I get the correct data?** Make sure that your selected features meet your criteria. For example, if you searched for rivers that are more than ten miles long, confirm that shorter rivers aren't included in your results.

» **Are the data appropriate for my purpose?** Check that the data's classification and attributes align with your project needs. Getting a soil map of Kansas only to find that the map is based on a Russian soil classification system doesn't do you much good if you're looking for classifications that align with U.S. agricultural standards.

» **Are the data on the right scale?** If you're interested in road networks for a city but search a database for an entire country, you face a good chance of missing key neighborhood streets.

» **Are the data timely?** Double-check that you're working with up-to-date information. Using outdated population figures from a decade ago might skew your analysis if more recent data are available.

» **Are the data complete?** Look for metadata — the "data about data" — to understand source details, coordinate systems, scale, and limitations in your data. When a dataset lacks metadata, treat it cautiously and always add metadata to your own work for reliability.

Errors happen, but handling them effectively keeps your analysis reliable. Here are some tips for managing errors and uncertainty:

» **Maintain attribute consistency:** Watch out for typos, missing values, or strange outliers in your attribute tables.

CHAPTER 9 **Finding Features in Vector Data** 149

» **Watch for spatial mismatches:** If datasets don't align perfectly (such as roads incorrectly overlapping rivers instead of crossing them at bridges or intersections), use snapping or topology tools to clean up these errors. Snapping ensures that roads meet at intersections, and topology tools fix overlaps or gaps between polygons.

» **Be up front about uncertainty in analysis:** Acknowledge any limitations in your data and analysis. For instance, if your buffer zones use estimated values, note that fact in your final report as well as in your metadata.

Metadata is one of your best friends! Check it for details on origins, scale, accuracy, and limitations. Always add metadata to datasets you create to ensure that they are trustworthy and reliable.

Validating and verifying your results is like proofreading your writing: Double-checking your GIS results ensures that you didn't misspell any features or forget to include critical data. By validating your results and managing errors, you can deliver trustworthy insights that your audience can rely on.

IN THIS CHAPTER

» Exploring why polygons matter in GIS

» Querying polygons effectively

» Understanding modern tools and workflows for data analysis

» Grouping data for actionable insights

Chapter **10**

Searching for Geographic Objects, Distributions, and Groups

Working with GIS is all about finding answers. Whether you're searching for a specific land parcel, trying to understand how a disease spreads, or analyzing where to build a new school, GIS gives you the tools to query, analyze, and act on your spatial data.

In this chapter, you explore one of the most powerful GIS capabilities: searching. You discover how to locate geographic objects (like rivers or roads), explore distributions (like forest areas or population clusters), and group features based on common traits or patterns. These techniques apply whether you're working with polygons, points, or lines, giving you the flexibility to tackle any project.

By mastering these search methods, you'll be able to identify patterns, uncover relationships, and generate insights that drive smarter decisions. So go ahead and dive into the practical tools and workflows that make GIS a real power tool.

Searching Polygons in a GIS

GIS was originally developed to conduct a land-use inventory to analyze forests, rivers, and mineral deposits as part of natural resources management. These large-area datasets, often represented by polygons, are called *distributions*. Because GIS is designed to help us understand spatial patterns, distributions are among the most important features of GIS analysis. Modern GIS tools allow you to map and analyze distributions of almost anything: people, wildlife, diseases, crime, land use, pollution, and more.

Because distributions represent areas rather than single locations (points) or connections and paths (lines), they are typically stored as polygons in GIS. Unlike points, which mark specific locations, or lines, which show connections between places, polygons define areas, enclosing space with clear boundaries. This feature makes them essential for analyzing spatial distributions and understanding patterns across large areas.

Understanding distributions is key to making informed decisions with GIS. (Check out Chapter 3 for a deeper dive into distributions and patterns.) Whether you're working with polygons that represent distributions of individual objects, such as animals or people, or broader area features such as land parcels or forests, polygons offer a wealth of information for analysis. With GIS, you can explore polygon properties, such as size, shape, and orientation, or dig into the data they contain, whether it's categorical, ranked, or numeric. Because polygons define areas, they allow for more complex queries, such as identifying overlapping regions, enclosed spaces, or adjacent areas, helping you analyze distributions in meaningful ways.

REMEMBER

Polygons aren't just shapes on a map; they have unique spatial relationships that set them apart from other features like points and lines. Polygons can touch, overlap, contain, surround, or be surrounded by other polygons. These relationships unlock powerful analysis opportunities with GIS. For example, you may want to find which forest polygons overlap a wildlife habitat or identify urban areas surrounded by industrial zones.

You can learn a lot with polygons. Polygons are versatile, packed with information, and essential for unlocking insights into your spatial data. The more you understand their properties and relationships, the more information you can extract using GIS. For example:

>> **Thematic content:** You can analyze what a polygon represents, such as types of vegetation, land use categories, or building footprints.

>> **Spatial metrics:** You can examine polygon properties like size, perimeter, area, or orientation.

>> **Changes over time:** You can track how a polygon's attributes or boundaries evolve, like monitoring the expansion of urban areas or deforestation trends.

Searching for the Right Objects

Searching for objects in GIS is like searching in real life: You need to know what you're looking for and why before you start. Whether you're looking for your car keys to drive to work, receipts to finish your taxes, or a phone number to send a text, every search has a purpose. The same holds true with GIS. You search for geographic objects like land parcels, city boundaries, or soil polygons because you have a goal: to understand patterns, solve a problem, or make a decision.

You may have a bunch of reasons for searching. Sure, sometimes you may just want to explore a map out of curiosity. For most of your searches, however, you'll have a specific goal in mind, and you look at a map to find information like the following:

>> **What categories are available:** For example, you may want to know what land-use categories the map displays.

>> **The size and shape of polygons:** Maybe you need to determine where a pine bark beetle infestation occurs in the forest.

>> **How polygons are arranged and located:** This information can help you identify, for example, what populations are at risk near potential hazardous materials spills.

Think of searching polygons as taking a journey to discover the "what" and "where" of your map. Approach your maps as questions waiting to be answered, and let those questions guide your exploration, measurements, and analyses.

Extracting specific information with attribute searches

Exploring geospatial data is fun (especially if you're a map nerd like me!). But it's not just fun for fun's sake. The real purpose is to extract specific information. For example, you might analyze the relationship between land-use types and elevation zones, or explore how or whether features relate to each other (such as crime hot spots and neighborhood demographics).

Modern GIS tools make attribute searches easy by combining user-friendly interfaces with Structured Query Language, or SQL (see Chapter 9 for more about SQL)). SQL is like a special language for asking questions about your data. It uses operators like AND, OR, and NOT, and comparative symbols such as =, >=, or < to build queries. Although you can write simple SQL expressions quickly, most GIS tools offer built-in interfaces to help you create searches without needing to memorize syntax. To see examples of these interfaces, refer to Chapter 9.

Sometimes, though, you need to edit or write SQL commands directly. Both ArcGIS Pro and QGIS give you the power to control your SQL, as follows:

» **In ArcGIS Pro**, toggle to view and edit the SQL syntax generated by the query builder, or write your SQL commands from scratch. (See Figure 10-1.)

» **In QGIS**, SQL commands are displayed natively as you click through the interface to create your query. You can also edit the SQL directly without needing to toggle any settings. (See Figure 10-1.)

Both of these tools enable you to view or edit SQL commands for greater control over your searches.

TIP

Misplaced parentheses or punctuation can mess up your queries. Thankfully, most modern tools let you validate or test your query before running it. When building queries, it's best to start small, test often, and refine as you go. Don't be afraid to experiment with writing your own SQL queries. Use the query builder in your favorite GIS software as a starting point and see what you can create!

Understanding polygon metrics

Polygons aren't just shapes — they're like little treasure troves of data waiting to be discovered. When you analyze polygons, you're not just looking at their outlines but uncovering the valuable nuggets within them. Most GIS software automatically calculates basic metrics like area and perimeter, but with a little effort, you can dig deeper to uncover even greater insights. Here are some examples of what you can analyze:

» **Shape descriptors:** Shape descriptors include characteristics like orientation (the angle a polygon faces), roundness (how circular or elongated it is), or axis lengths (dimensions along its longest and shortest sides). Following are some examples of how shape descriptors can be helpful:

• **Orientation:** Determine the best parcels for solar panels or wind farms.

• **Roundness:** Identify compact urban lots versus irregular agricultural fields.

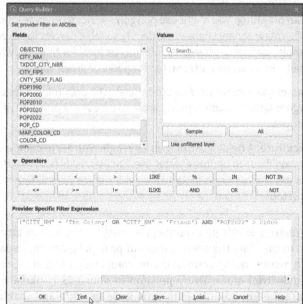

FIGURE 10-1: The ArcGIS Pro Select By Attributes interface (top) and the Query Builder (bottom).

CHAPTER 10 Searching for Geographic Objects, Distributions, and Groups   155

- **Axis lengths:** Highlight parcels suitable for linear infrastructure, such as roads and pipelines.

» **Distance metrics:** These metrics calculate how far polygons are from each other and can help you answer questions like

- Are wildlife habitats fragmented, or do they form a connected corridor?
- How far are industrial zones from residential neighborhoods, and are they at safe distances?
- Which retail stores have overlapping service areas, and are there gaps between service areas?

Understanding polygon metrics helps you make informed decisions across a variety of applications, such as the following:

» **Environmental studies:** You can calculate the size and shape of forest patches to assess their biodiversity potential. Large, contiguous polygons are better for wildlife corridors, whereas smaller, fragmented polygons may indicate habitat degradation.

» **Urban planning:** You might use polygon metrics to analyze lot sizes in a city to determine zoning changes or identify suitable parcels for development. For example, compact, rectangular lots may be ideal for housing, whereas irregular lots may be better suited for parks or green spaces.

» **Disaster management:** You can calculate the proximity of residential polygons (lots) to wildfire-prone areas to assess risk and plan mitigation strategies.

» **Agriculture:** A helpful use of metrics in agriculture is to compare shapes of irrigation fields to optimize water use. Circular fields tend to indicate center-pivot systems, whereas rectangular fields may require different irrigation setups.

TIP

Polygon metrics aren't just numbers — they're the key to meaningful analysis. Use them to calculate the compactness of park polygons for expansion prioritization, or to assess zoning compliance by comparing the distance between residential and industrial areas. Every polygon in your GIS is like a treasure chest; its metrics are the key to unlocking actionable insights and making smarter decisions. Next time you fire up your GIS, don't just admire the polygons but also explore their data, measure their metrics, and uncover the gems waiting inside!

Analyzing point distributions

At first glance, scattered points on a map may seem random. But with GIS, you can turn them into meaningful patterns and relevant results that help you make sense of data. Tools like hot-spot analysis and kernel-density estimation help you uncover clusters and areas of high concentration. *Hot-spot analysis* typically outputs polygons that highlight statistically significant clusters, and *kernel-density estimation* creates a smooth raster surface showing density patterns. Both can reveal deeper relationships and trends when combined with other data.

For example, the heat map from the Frisco, Texas, Police Department Crime Reports Dashboard, shown in Figure 10-2, reveals the intensity of criminal activity across the city over a 30-day period. The dark spots, especially noticeable in the southern end of the city, indicate high crime density, which helps law enforcement focus their resources. This is a great example of how point data can reveal trends that might not be obvious at first glance.

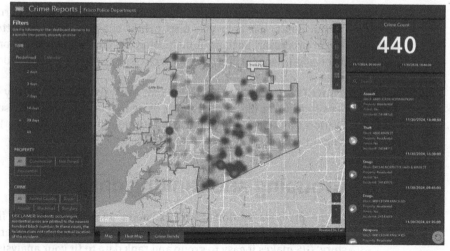

FIGURE 10-2: The Crime Reports dashboard by the City of Frisco, Texas, uses a heat map to indicate high-density crime areas.

Map from Crime Reports, Frisco Police Department

But the usefulness of point distribution techniques doesn't stop at crime mapping. These techniques are incredibly versatile, letting you visualize and analyze data across a wide range of fields. Here are some real-world scenarios of how point distributions can uncover valuable insights:

>> **Map species habitats:** Use GPS-tagged wildlife sightings to create polygons representing core habitat areas. Compare these habitat polygons to vegetation patterns or human activity to identify factors influencing the species' distribution.

CHAPTER 10 **Searching for Geographic Objects, Distributions, and Groups**   157

» **Analyze voter turnout:** Map polling station attendance as points and generate polygons showing areas with high or low turnout. Overlay these polygons with demographic data to study trends or disparities.

» **Study disease breakouts:** Map reported cases of measles and create polygons around high-density areas. Compare these polygons with factors like population density, vaccination rates, and proximity to schools or healthcare facilities to identify potential patterns and risks.

» **Conduct retail site analysis:** Map customer addresses as points and generate service area polygons for each store based on distances or drive times. Use the resulting polygons to find overlaps, service gaps, or underserved neighborhoods.

» **Plan emergency responses:** Map the locations of emergency calls and create polygons outlining high-demand areas. Compare these areas to the locations of fire stations or hospitals to determine where additional resources are needed.

Point distribution analysis tools like hot-spot analysis and kernel-density estimation don't just help you map points but also uncover patterns that reflect deeper processes, such as human behavior, environmental factors, or infrastructure design. Recognizing these patterns is key to making smarter decisions with GIS.

In ArcGIS Pro, the Optimized Hot Spot Analysis tool can identify clusters of high-density features like crime reports or dense housing areas. These types of identification can also be done in QGIS, but you'll need to add the Hotspot Analysis plug-in.

Grouping and ranking data

As the previous section describes, point analysis helps you find clusters and trends. Sometimes, however, you just need to rearrange your data to tell the right story. GIS makes it easy to group and rank data to fit your analysis goals. Grouping data simplifies the analysis by reducing complexity, and ranking can help you prioritize areas for action based on specific criteria. For example:

» **Grouping categories:** Grouping similar features into broader categories can simplify your analysis and make it easier to compare data. For example, grouping tree types like Pine, Oak, and Birch into a single "Forest" category can make it easier to analyze land across federal and private ownership. Figure 10-3 shows this grouping in action. This kind of grouping can be useful when you need to simplify complex datasets or align them with broader analysis goals.

> » **Changing data measurement scales:** Sometimes, raw categories need to be converted into rankings or scores to fit your analysis. For example, if you're studying soil erosion risk, you can assign numeric values to land cover types where "bare soil" = 1 (most susceptible) and "grassland" = 5 (least susceptible). This approach helps you compare different categories on a single scale and prioritize areas for intervention.

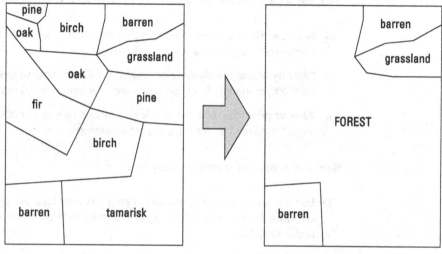

FIGURE 10-3: You can aggregate tree species data into a category called Forest.

© John Wiley & Sons, Inc.

REMEMBER

When you group or rank data, you're not changing the original dataset but rather rearranging it to answer your specific question. Think of reshuffling a deck of cards: The cards themselves don't change, but the way they're organized can reveal new patterns.

Locating Map Features

GIS offers a wide range of tools to help you search for polygons based on their attributes, shapes, locations, or relationships to other features. Whether you're looking for a specific type of land use, identifying clustered polygons, or combining search criteria, these methods help you discover valuable insights in your data. In this section, you get to know five practical ways for locating features.

REMEMBER

These methods provide a comprehensive framework for locating features in GIS. By learning and combining these approaches, you can gain a deeper understanding and solve complex spatial questions, whether you're planning urban developments, studying ecological patterns, or mapping public health data.

CHAPTER 10 **Searching for Geographic Objects, Distributions, and Groups** 159

Searching by attributes

Attributes answer the "what" question for GIS data. Want to know which parcels are zoned for residential use, or how many forest polygons have trees older than 50 years? Searching by attributes gives you a straightforward way to zero in on the features you care about. Whether you're categorizing, filtering, or prioritizing, this method is as versatile as it is essential.

Here are some examples of what you can do in an attributes search:

» **Find specific categories:** Find polygons classified by a specific type, such as commercial zones, wetlands, or national parks

» **Filter by numeric values:** Select polygons with property values below $250,000, or identify forest polygons with tree ages over 50 years

» **Rank or prioritize:** Search for polygons ranked by levels of risk, such as "high," "medium," or "low" in a hazard assessment

Here's how you can search by attributes:

» **Use a SQL query or search tools in your GIS interface.** For example, you can write the following SQL query to search for residential parcels valued under $250,000:

```
SELECT * FROM parcels WHERE zoning = 'Residential' AND
    value < 250000
```

» **Filter through a map legend.** Filtering through a legend is one of the easiest ways to search for specific categories and is especially useful for nontechnical users. Many interactive maps allow you to toggle visibility for categories directly from the legend. For example, if you're looking to open a pet grooming shop, you can use an interactive tool like the Census Business Builder, shown in Figure 10-4, to identify potential areas to locate your business. This web map enables you to filter and analyze data by topics like economic activity and demographics.

Explore the Census Business Builder tool at https://cbb.census.gov/ to practice filtering polygons based on economic data.

Searching by shape and size

Polygons may look simple, but their shapes and sizes can hold the key to solving your spatial problems. From finding large, open parcels for solar farms to pinpointing compact urban lots, searching by geometric properties — size, shape, and orientation — helps you dig deeper into your data.

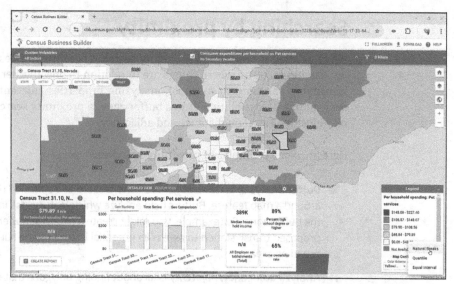

FIGURE 10-4:
The Census Business Builder.

Map from U.S. Census Bureau

Here are some examples of searching by shape or size:

>> **Search by size:** Find polygons larger or smaller than a specific area – like searching for parcels larger than 10 acres or smaller than 1 acre.

>> **Analyze shapes:** Identify polygons with specific ratios, such as compact or elongated shapes, or identify circular polygons versus irregular ones.

>> **Search by orientation:** To find areas ideal for solar panel installations, you can find polygons aligned north-south.

Here are ways to perform such searches:

>> **Use geometry calculators:** Both QGIS and ArcGIS Pro have built-in tools to compute polygon areas, perimeters, and shape descriptors. You can then filter polygons based on these attributes.

>> **Combine multiple filters:** If you're planning a wind farm, you need large, flat, and open land. To find ideal locations, you would follow these steps:

1. Use a geometry tool to filter for polygons larger than 100 acres.

2. Select compact polygons from the filtered group.

3. Refine your results by finding polygons near key infrastructure elements, like roads or power lines.

CHAPTER 10 Searching for Geographic Objects, Distributions, and Groups

Searching by proximity

When location is everything, proximity searches become your best friend. *Proximity* searches focus on where things are in relation to one another. Whether you're identifying parcels near a school, analyzing properties adjacent to wetlands, or determining accessibility within a buffer zone, a proximity search helps you map relationships based on distance and adjacency.

Here are some examples of proximity searches:

» **Buffer analysis:** In GIS, a *buffer* is a zone created at a specified distance around a map feature. Buffers help analyze which features fall inside or near the buffer zone. For example, you might create a two-mile buffer around a proposed transit station to assess its potential impact on the surrounding area.

» **Analyze adjacency:** *Adjacency* is all about finding polygons that share borders. For example, you might need information about properties that border (are adjacent to) a park, wetland, or other feature of interest.

» **Search for nearby polygons:** If your study is limited to a particular area, you may need to identify parcels within city limits or regions within a specific distance of a highway.

Here are ways to perform proximity searches:

» **Apply buffer tools:** Most GIS platforms let you create buffer zones. For instance, if you're planning a new hospital, you might follow these steps:

1. Create a 10-mile buffer around potential sites.

2. Analyze the demographics of populations within each buffer.

3. Choose the site serving the largest population in need.

» **Use adjacency tools:** In ArcGIS Pro, use the Select by Location tool to find polygons touching other features. For example, for a floodplain analysis, find all properties adjacent to a river.

Searching by groups and clusters

Individual features often tell only part of the story. Many times, though, the patterns they form as groups or clusters reveal the full story. Clustering features—such as wildlife habitats, agriculture fields, or densely packed neighborhoods—helps you see the bigger picture and identify trends that might not be obvious at the individual feature level.

Here are some examples of searching by groups and clusters:

- **Group by density:** Identify areas where features are densely packed, such as regions with intersecting wildlife habitats or contiguous agricultural fields forming a cohesive farming zone.
- **Cluster polygons with shared characteristics:** Group polygons that share common attributes or spatial relationships, such as neighboring parcels zoned for residential use or forest stands with similar trees.
- **Analyze connected features:** Find polygons that are spatially connected, such as floodplains with adjacent wetlands or industrial zones located next to residential neighborhoods.

Here are ways to perform these kinds of searches:

- **Use clustering on point data to inform polygons:** Tools for finding groups in scattered data, like DBSCAN (Density-Based Spatial Clustering of Applications with Noise) in QGIS, or ArcGIS Pro's Density-based Clustering tool, can help identify point clusters, such as wildlife sightings or reported disease cases. After you identify them, these clusters can be converted into polygons that represent broader zones, like animal habitats or outbreak areas.
- **Aggregate attributes into new polygons.** Combine adjacent polygons based on shared characteristics or spatial patterns. For example:
  - **Develop wildlife conservation zones:** Group neighboring polygons with similar ecosystems to delineate areas critical for wildlife habitat.
  - **Build agricultural zones:** Combine farmland polygons growing the same crop to analyze irrigation needs and pesticide use.
- **Analyze adjacency:** Use GIS tools like *spatial joins* (which join attributes based on location) to find polygons that are spatially connected or share characteristics. For example, identify flood-prone areas by grouping polygons with similar slope, soil type, and proximity to rivers.

Combining multiple search methods

Why settle for one search method when you can combine them? Real-world questions often demand layered analyses, like finding affordable housing within two miles of public transit. By blending attributes, proximity, and spatial metrics, you can tackle even the most complex GIS problems with confidence.

Here are some examples of combining multiple search methods:

» **Filter by multiple attributes:** For example, find convenience stores within a half mile of schools.

» **Add spatial metrics:** After searching for vacant land parcels for a solar project, you can narrow your results by selecting compact parcels or those oriented north-south.

Here's how you can perform these types of searches:

» **Write complex SQL queries:** For example, combine your criteria into a SQL search tailored to your specific need, which in this example is finding affordable housing within two miles of transit stations:

```
SELECT * FROM parcels
WHERE zoning = 'residential'
AND value < [Affordable Housing Threshold]
AND ST_Distance(geometry, transit_station) < 2
```

» **Save your query results:** Most GIS platforms let you save your query results as a new layer so that you can revisit or build on your results without starting over.

Defining the Groups You Want to Find

Grouping features in GIS is a common method for uncovering meaningful patterns, whether you're working with points, lines, or polygons. Although this chapter mostly highlights polygons, many of the methods covered apply to all three feature types, each with its own unique spatial characteristics that influence how you use them for analyses. But wait, there's more! The next sections look at three practical ways to group your data — whether you're working with points, lines, or polygons — and extract insights.

Grouping by common properties

Grouping features by shared attributes is one of the most straightforward and versatile GIS grouping methods. To find groups of points, lines, or polygons, you start by finding what those features have in common. Think of it like sorting your music playlist by genre, artist, or decade, but now you're working with GIS data. Here's how you can group features by what they have in common:

- **Category:** Select all features of the same type, such as fast-food restaurants (points), major highways (lines), or forested areas (polygons).

- **Rank:** Sort features by a measurable quality, like cities by population size (points), roads by traffic volume (lines), or land areas by flood risk level (polygons).

- **Size:** Find high-rise buildings taller than 200 feet (points), roads longer than five miles (lines), or parks larger than 10 acres (polygons).

These methods let you focus on features most relevant to your goals, regardless of their type.

TIP

Many GIS tools let you filter data easily by clicking to select fields and values or by writing simple queries (refer to Figure 10-1). For example, to find parks larger than 20 acres within a city, you can write a query like this:

```
SELECT * FROM parks
WHERE area > 20
AND CITY = 'Alexandria'
```

Building queries might seem intimidating, but it's simply telling your GIS software what you're looking for. Think of it as giving your map a "to do" list. Give it a try with your own data!

Grouping by location and patterns

Sometimes, the story isn't in the attributes — it's in the map. GIS can help you find patterns like the following in how features are arranged:

- **Clusters:** Identify dense groupings, like overlapping wildfire reports (points), frequently intersecting railroads (lines), or neighborhoods with tightly packed housing (polygons).

- **Alignment:** Spot trends such as utility lines following a river (lines) or fields aligned with irrigation ditches (polygons).

- **Proximity:** Highlight features close to each other, like bus stops within walking distance of schools (points) or land parcels near highways (polygons).

Using tools like clustering and proximity analysis helps you see how features relate to one another, no matter their type. Because polygons have spatial relationships that points and lines don't, they are especially important when analyzing how areas connect and influence each other. Unlike points or lines, polygons show how areas overlap, contain, or align with one another, such as whether a flood zone overlaps multiple zoning districts or if agricultural areas cover specific soil types.

Grouping by what you already know

When it comes to grouping and organizing your data, sometimes the best place to start is with what you already know about your data or the area you're studying. This prior knowledge, whether it relates to existing spatial relationships, patterns, or data layers, can help you uncover new information and piece together what you don't know.

Whether you're working with points, lines, or polygons, the following methods can help you effectively group and analyze your data:

» **Overlay existing data:** Combine known data layers to reveal relationships. For example, compare emergency call locations (points) to fire station coverage areas (polygons).

» **Spot familiar patterns:** Use your understanding of features to form hypotheses. For example, if you know that flood-prone areas (polygons) often align with low-lying road networks (lines), you can investigate whether some roads are at greater risk during heavy rains.

» **Fill the gaps:** When your data doesn't align as expected, such as crime hot spots (points) being located outside neighborhood boundaries (polygons), you may discover missing information or new trends.

REMEMBER

Your GIS analysis starts with what you know, but the most valuable insights emerge from exploring what you don't know. By combining existing data, questioning gaps, and refining your approach, you can uncover insights that help you make smarter decisions.

4 Analyzing Geographic Patterns

IN THIS PART . . .

Discover how to use GIS to measure distances, from basic straight-line calculations to real-world routes that factor in time, cost, and obstacles.

Explore statistical surfaces to analyze geographic patterns and predict missing values in your data.

Use topographic surfaces to understand how to trace where water flows and reveal changes in the landscape.

Find out how to optimize movement and connectivity in networks like roads, utilities, and rivers while factoring in real-world constraints.

Discover how to use overlay techniques to compare map layers and reveal spatial relationships.

Master map algebra and cartographic modeling to combine datasets, automate workflows, and perform spatial analysis.

> **IN THIS CHAPTER**
>
> » Measuring absolute distance
>
> » Exploring relative distances and their practical uses
>
> » Understanding functional distance impact decisions

Chapter 11
Measuring Distance

Maps do more than just show you where things are; they help you measure and understand the world around you. Ever wondered how far you are from a good coffee shop or how much land your backyard covers? GIS takes the guesswork out of measuring distances, giving you tools to calculate everything from the shortest path to a trailhead to the size of a protected wetland area.

One of the greatest strengths of GIS is its speed and precision in measuring geographic features. Whether you're comparing routes to find the shortest commute, assessing how close a new development is to a protected wetland, or pinpointing the best slopes for skiing, GIS delivers results quickly and accurately.

In this chapter, you explore how GIS handles measurements of all kinds. It starts with the basics, like straight-line distances and area calculations, and then dives into advanced concepts, such as relative measurements (how close is close enough?) and functional distance (what does it cost to get there?). In this chapter, you see how GIS can do more than just measure; it can also transform raw data into actionable insights!

Taking Absolute Measurement

GIS is great at measuring distances and areas quickly and accurately. When you enter data into a GIS, the software records key details about points, lines, and polygons, including their X and Y coordinates and, for raster data, the size of grid cells. These built-in measurements form the foundation for analyzing spatial relationships, helping you answer questions like "How far?", "How big?", and "What's nearby?"

But not all measurements are created equal. For example, measuring the distance to the nearest grocery store doesn't require taking into account Earth's curvature, but planning a shipping route from Los Angeles to Sydney absolutely does. The difference lies in whether your analysis requires the use of flat measurements — like measuring the straight lines of a countertop — or spherical measurements that trace the curve of the globe. By understanding when to use flat or spherical measurements, you can ensure that your results stay relevant and accurate.

Fortunately, GIS simplifies these decisions by adapting to the type of surface involved, whether you're working with flat maps, spherical surfaces, or intricate networks like roads and rivers. Each scenario requires precision, and GIS gives you that through the use of map projections. Map projections, which I describe in Chapter 2, are important for maintaining measurement accuracy. However, the method used for measuring is also important.

Finding the shortest straight-line path

The simplest way to measure distance is "as the crow flies," of course! This method is also known as the straight-line path or Euclidean distance. *Euclidean distance* is the shortest distance between any two points in a straight line, as if you were drawing a line between the points with a ruler. This method makes two assumptions:

>> No obstacles block the path.

>> You don't need to follow roads, trails, or other routes.

This type of measurement is perfect for getting a quick sense of the distance between two places. Say you're planning a hiking trail through a forest or checking how far wildfire embers might jump. Determining straight-line distance gives you an initial snapshot of these situations. The shortest path measurement is also handy for estimating how close two potential new business locations are to each other, or for scoping out the shortest path for a new road alignment.

Measuring on a flat surface

If your map represents a flat surface, finding the shortest distance between two points is simple geometry. GIS uses the *Pythagorean Theorem* to calculate distances, which you may remember from geometry class as ($c^2 = a^2 + b^2$). To see how it works, take a look at the right triangle in Figure 11-1, where:

>> The two legs represent the horizontal (a) and vertical (b) distances between the points A and B.

>> The *hypotenuse* is the straight-line Euclidean distance (c) you're measuring.

The Pythagorean Theorem may be as ancient as the Greek mathematician, Pythagoras, that it's named after, but your modern GIS uses it to compute distances in 2D, flat-surface maps, making quick work of problems that once required lengthy calculations.

REMEMBER

Although GIS handles the math for you behind the scenes, understanding measurement basics empowers you to use the tools more effectively, troubleshoot issues, and think critically about spatial problems, whether you're deciding where to lay a new water line or measuring property boundaries.

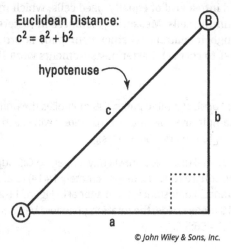

FIGURE 11-1: A right triangle using the Pythagorean Theorem.

© John Wiley & Sons, Inc.

Measuring on a spherical Earth

Unlike flat maps, the Earth is a sphere (well, almost a sphere, as explained in Chapter 2). For longer distances, especially across continents or oceans, GIS uses *great circle distance*, which is the shortest path between two points on a sphere. Picture the great circle distance as tracing an arc on a globe rather than drawing a straight line on a 2D or flat map.

CHAPTER 11 **Measuring Distance** 171

The great circle distance method accounts for Earth's curvature, which is critical for accurate measurements over long distances, like flights from Los Angeles to New York that follow an arc across the globe. To use great circle distances, your GIS relies on projections, which I describe in Chapter 2.

The best part? GIS handles the calculation for you, so you can focus on what distance means for your analysis, whether you're planning shipping routes or calculating travel times.

TIP

When analyzing long distances, such as transcontinental routes or international flight paths, check the GIS tool settings or documentation to ensure that the tool is using geodesic calculations for great circle distance. Using a geographic coordinate system (GCS) like NAD83 or GRS80 and enabling geodesic options in distance tools ensures accuracy for large-scale analyses.

Measuring distances in grid cells

Grid cell measurements are useful when using raster data to model wildlife movements, calculate slope, or map flood risks. For example, flood modeling uses raster-based measurements to predict water spread across a landscape. (Refer to Chapter 5 for more about raster models and their structure.)

Rasters divide the world into a grid of equally sized cells, which makes measuring distance a matter of counting cells. Measuring distances in rasters is conceptually straightforward, although it matters whether your path is *orthogonal* (edge to edge) or *diagonal* (corner to corner). Raster measurements work like this (and see Figure 11-2):

» **Orthogonal paths:** For straight-line paths, GIS multiplies the number of cells by their size. For example, five cells that are 10 meters wide each is a distance of 50 meters (5 cells x 10 meters = 50 meters).

» **Diagonal paths:** Diagonal distances are slightly longer, so GIS adjusts the calculation using the square root of 2 (approximately 1.414) to account for the longer diagonal distance. For instance, the diagonal in Figure 11-2 has four cells. If each cell is 10 meters wide, the total length would be 56.57 meters (4 cells x 10 meters X $\sqrt{2}$ = 56.57).

Measuring Manhattan distance

Ever walked through parts of a grid-like city, such as Manhattan, where you follow the streets rather than take a direct path? That's where Manhattan distance comes in. Instead of measuring a straight line, *Manhattan distance* adds up your journey's horizontal and vertical segments.

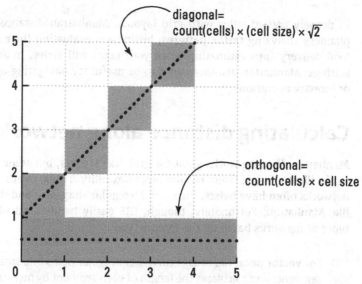

FIGURE 11-2: Measuring diagonal and orthogonal distance in grid systems.

© John Wiley & Sons, Inc.

For example, take a look at Figure 11-3 and imagine you're using GIS to plan delivery routes for a local business. Your GIS will calculate the distance to get from the business at point A to the customer at point B by summing up the city blocks (the horizontal and vertical segments).

Although it's longer than the straight-line Euclidean distance, Manhattan distance reflects the reality of navigating a city grid. For example, when your food delivery app estimates the time that your late snack will arrive at your door, it's factoring in Manhattan distance — navigating your city block by block.

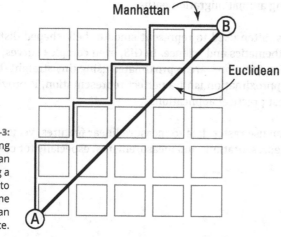

FIGURE 11-3: Measuring Manhattan distance along a grid compared to straight-line Euclidean distance.

© John Wiley & Sons, Inc.

CHAPTER 11 Measuring Distance   173

In densely packed cities with grid layouts, Manhattan distance is essential for planners analyzing traffic patterns, businesses evaluating their service areas, or food delivery apps estimating when your pizza will arrive. In addition to urban settings, Manhattan distance can also be useful for navigating agricultural grids or forestry operations.

Calculating distance along networks

Manhattan distance works great for grid-like layouts, but some real-world networks like trails and rivers require more flexibility. Trails, rivers, roads, and other networks often have twists, turns, and irregular shapes — and they look nothing like Manhattan. Fortunately, though, GIS easily handles calculations along all types of networks based on the feature type:

- » **For vector data:** GIS breaks curved or irregular paths into small line segments and calculates the length of each segment by measuring the coordinates of its endpoints. It then adds up all the segments for the total distance, capturing the shape of the path, whether it's a winding river or scenic hiking trail.

- » **For raster data:** GIS counts grid cells, just like in the previous section, and adjusts for diagonal paths, as shown in Figure 11-2. Although less precise than vector data, raster data works well for modeling broader patterns like wildlife migration.

Imagine you're planning an evacuation route for a natural disaster. GIS calculates not just the shortest path but also the fastest one, considering road closures and congestion. Calculating distances along networks is crucial for applications like finding the shortest route for emergency vehicles, mapping public transit systems, or studying animal migration paths.

Gaussian curves, often used to represent smooth, bell-shaped distributions, are common in mathematics and science. In GIS, even complex curves, such as those based on Gaussian functions, are approximated using tiny straight-line segments. Although this approximation isn't a perfect representation, it provides sufficient accuracy for most practical applications.

Although you can use raster data to measure linear features, vector data provides more accurate representations and measurements, especially for curving or irregular paths.

Establishing Relative Measurements

Imagine that you're deciding where to build a new coffee shop. You don't need the exact distance of every competitor; you just need to know whether they're close enough to pose a threat. Relative measurement helps you understand spatial relationships without needing precise measurements.

Relative measurement in GIS helps answer questions like "How close is close enough?", "What's inside what?", and "How isolated is something?" GIS provides relative measurement tools to answer these questions and quantify spatial relationships. For example, GIS can help you analyze how parcels interact with a nearby lake, assess whether patches of invasive plants are spreading too close to each other, or determine whether a park is fully surrounded by urban sprawl.

Relative measurement comes in different forms, depending on the type of spatial relationship you're analyzing. Sometimes, relative measurement is about adjacency and nearness, such as determining whether two features share a boundary or are close enough to be considered nearby each other. Other times, relative measurement relates to separation and isolation, such as determining whether features are spatially disconnected or part of the same network. And in some cases, the focus of your measurement is on containment and surroundedness, where one feature is inside or encircled by another.

GIS provides relative measurement tools to evaluate these types of spatial relationships, helping you answer questions like these in the following categories:

- **Adjacency and nearness:** Which land parcels directly border a lake or fall within a certain distance?
- **Separation and isolation:** Are two patches of invasive plants close enough to spread between them, or are they effectively isolated?
- **Containment and surroundedness:** Is a property located within a flood zone, or is a park completely surrounded by urban development?

Understanding these spatial relationships allows GIS analysts to make informed decisions on topics like planning new developments, managing ecological threats, or assessing accessibility in urban spaces.

Adjacency and nearness

When it comes to relative measurement, adjacency is one of the most important concepts to understand. Adjacency occurs when a feature is so close to another that they share a boundary or touch, with the distance essentially zero. GIS makes identifying adjacency and defining what counts as "near" both precise and straightforward.

Take a look at Figure 11-4 and imagine that you're evaluating land parcels near a lake. Follow these steps to determine adjacency or nearness:

1. **Create a quarter-mile buffer around the lake, as shown in Figure 11-4, by selecting the lake polygon and specifying the exact distance you want to analyze.**

2. **With the buffer established, use a tool like Select By Location (available in both ArcGIS Pro or QGIS) to highlight polygons that touch or overlap the buffer.**

   The selected polygons shown in Figure 11-4 are considered "adjacent to" or "near" the lake.

REMEMBER

Adjacency isn't just about buffers; it's also integral to topological models. (Chapter 5 explains the various data models.) With topology, GIS can identify polygons that share edges or corners without needing to create buffers, allowing for highly efficient analysis. Whether you're a real estate developer evaluating parcels near a lake or a conservationist analyzing habitat connectivity, adjacency helps you determine relationships between features.

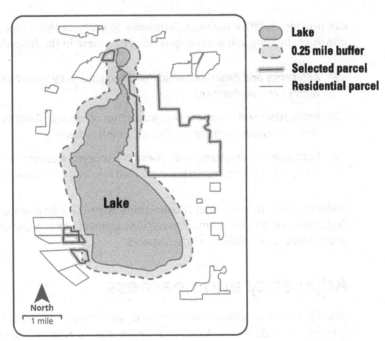

FIGURE 11-4: A lake with a quarter-mile buffer and parcels selected that fall within or touch the buffer.

© John Wiley & Sons, Inc.

Separation and isolation

Imagine that you're an ecologist tackling the spread of invasive plants like Buffel grass. This hardy species spreads rapidly by producing a large amount of seeds, which can be carried by wind, water, or animals to colonize new areas. To manage its spread, you need to determine whether patches of Buffel grass (*Cenchrus ciliaris*) are close enough for seeds to disperse between them. That's where GIS comes in, helping you to measure just how connected (or separated) these patches are and offering valuable insights for targeted control efforts.

Here's how GIS can help you:

>> **Pinpoint the shortest connections:** Calculate the shortest distances between the edges of Buffel grass patches. These calculations help to identify areas where seed dispersal is most likely to occur. For example, patches that are only a few meters apart can create corridors for further spread.

>> **Focus on centroids:** Measure distances between centroids (the geometric centers) of the patches to understand their overall layout. This focus on centroids helps you spot isolated patches that may not yet contribute to the spread, allowing you to prioritize control measures.

By analyzing connectivity, GIS gives you valuable information for managing Buffel grass. If two patches are too close for comfort, breaking their connection may slow the spread. With GIS, you're not just observing the problem; you're also creating a plan to solve it.

REMEMBER

GIS helps you analyze connectivity and accessibility, whether you're assessing wildlife corridors, planning green spaces in urban areas, or evaluating the spread of invasive species.

Containment and surroundedness

Sometimes the question isn't "how far?" but rather "what's inside?" (containment) or "what's around?" (surroundedness). *Containment* tells you what's inside, such as a property within a flood zone. *Surroundedness* looks at whether a feature is encircled or enclosed by another feature, such as whether a park is hemmed in by residential neighborhoods. GIS makes these assessments simple by analyzing the relationship between features.

GIS determines containment by checking whether an entire feature — such as a point, line, or polygon — falls within the boundaries of another feature, often called the containment polygon. Here are some examples of how you might use containment polygons:

- **Flood zone:** Is your home inside a 100-year floodplain?
- **Wildlife habitat:** Does a species live in this protected area?
- **Voting district:** Which district covers your address?

GIS tools can also compare multiple layers, such as precinct boundaries and street maps, to answer containment questions. It can do so even for partially contained features, such as a road running partly through a park.

Surroundedness, on the other hand, goes a step beyond containment. It focuses on whether a feature is encircled by another, even if it isn't fully contained. For example:

- A park can be surrounded by residential neighborhoods without being "inside" any of them.
- A wildlife preserve surrounded by urban sprawl may not be contained by the city limits, but it's fully encircled, highlighting potential connectivity issues.
- An island is surrounded by a lake, but the water doesn't contain the island — it surrounds it.

The distinction between containment and surroundedness is critical when analyzing spatial relationships that involve proximity, encirclement, or visual dominance. Analyzing surroundedness might involve identifying parks entirely hemmed in by urban development, highlighting potential issues or ecological isolation.

Measuring Functional Distance

Ever wonder why Google Maps gives you an estimated time instead of just the distance? That's because distance isn't always about how far away a location is; it's also about how hard it is to get there. *Functional distance* refers to the costs, effort, or time it takes to get somewhere. It's the difference between "How many miles is it to the highway?" and "How long will it take me to drive there?" It also helps you plan smarter routes.

Functional distance captures the realities of movement, factoring in things like time, money, fuel use, and even emotional effort. For example, that shortcut through the neighborhood might be shorter, but with school zones and stop signs at every intersection, that route may involve some aggravation.

Here are some factors that define functional distance:

- **Fuel and energy use:** How much gas does your commute drive consume?
- **Cost:** Does one route have tolls or parking fees?
- **Time:** Are there obstacles, speed limits, or other factors that will take more time on one route over another?
- **Effort:** Is the terrain rugged or smooth? How well maintained is the road?
- **Stress level:** Are traffic jams or steep, winding mountain roads that add mental strain?

Functional distance isn't about precision; it's about practical decision-making.

Navigating non-uniform surfaces (anisotropy)

If you've ever walked on a sandy beach, hiked a steep trail, or bicycled into a headwind, you've experienced *anisotropy*, which is a fancy term for uneven travel conditions that you can drop at dinner parties to impress (or confuse) your friends. It's the opposite of *isotropy*, which means a perfectly smooth surface that enables equally easy travel in all directions. Isotropy is purely theoretical; it doesn't exist in nature but gives you a standard to compare surfaces.

Imagine traveling across an isotropic field, with every step requiring the same amount of energy no matter which way you go. Real life isn't like that, of course; it's anisotropic. Roads, trails, and natural landscapes introduce resistance, or what GIS gurus call "friction." Figure 11-5 shows that areas with lower friction allow for faster and easier travel. Factors like slope, surface type, and obstacles all create varying degrees of friction, influencing how far you can go and how quickly you can get there. Each factor contributes to anisotropy as follows:

- **Slope:** Steeper slopes mean greater effort for vehicles, cyclists, and hikers.
- **Surface type:** Smooth pavement is a dream compared to soft sand or slippery mud.
- **Surface features or obstacles:** Dense vegetation or rocky terrain can slow you down dramatically.

CHAPTER 11 **Measuring Distance** 179

REMEMBER

Aside from being a fun term to drop at parties, anisotropy is an important concept to grasp in GIS. It highlights the real-world factors that make travel conditions uneven, like steep hills or rough terrain. By modeling these anisotropic challenges, GIS incorporates terrain-based travel costs into its analysis. Your results, therefore, aren't just theoretical but are grounded in the practical realities of movement. Whether you're planning disaster evacuations, laying out powerlines, or assessing how terrain impacts agricultural irrigation, understanding anisotropy provides you with realistic and actionable insights.

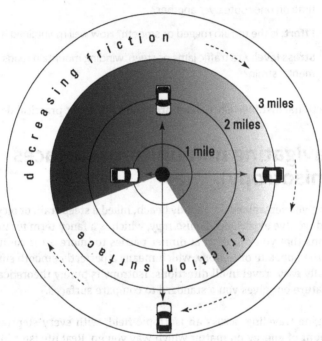

FIGURE 11-5: Rough (high-friction) surface travel versus smooth (low-friction) surface travel.

© John Wiley & Sons, Inc.

Considering the intangibles

Not all factors affecting functional distance are physical. Sometimes the journey is less about the miles and more about the experience, or the *intangibles* — those subtle, nonphysical things like how it feels:

» **Scenic routes:** A longer road through the mountains may be more appealing than a faster freeway.

» **Mental stress:** Heavy traffic or a narrow, windy, cliffside road may make a shorter trip feel longer.

» **Personal preferences:** I may prioritize avoiding highways and toll roads, whereas you may opt for the smoothest ride.

REMEMBER

Functional distance goes beyond the numbers and reflects the human experience of travel. GIS makes it possible to incorporate these intangible factors into your analysis by combining subjective or qualitative data with traditional data, such as gathering user feedback about city bike paths or factoring scenic route preferences into driving tour designs. Understanding functional distance helps you create solutions that balance practicality with the user experience.

Creating the functional surface

GIS tackles functional distance by creating a *friction surface*, which is a map layer that assigns resistance values to different areas. *High friction* means that the route is tough to travel (think dense forests or steep hills), whereas *low friction* means smooth sailing (like highways or open fields). The friction surface layer becomes the foundation for modeling movement across landscapes.

GIS assigns friction in the following ways:

» **Physical resistance:** Features like rivers, cliffs, and roads are assigned values based on travel difficulty.

» **Human preferences:** Intangible factors, like avoiding heavy traffic, can also be incorporated.

» **Scalar values:** Friction ranges from 0 (no friction) to 10 (impassable), thereby allowing for flexibility in your models.

GIS combines these friction values to create a realistic picture of travel challenges across a landscape (refer to Figure 11-5). Combining these values ensures that your analysis reflects actual conditions instead of just straight-line distances, helping you to identify routes, barriers, and areas requiring extra effort.

Calculating the functional distance

After creating your friction surface, GIS combines it with distance equations to estimate travel costs. Here are some common methods for calculating functional distance:

» **Grid-based calculations:** Raster models divide the map into cells. Each step between cells adds distance and friction.

CHAPTER 11 **Measuring Distance** 181

- » **Cumulative totals:** As GIS calculates distances, it adjusts for friction at each step. Higher friction means higher travel costs.
- » **Directional differences:** Uphill travel adds more resistance than downhill, making routes more realistic.

For example, to calculate the time it takes to cross a forest, a GIS model may assign a friction value of 3 to a dirt trail, 7 to dense vegetation, or 0 to a paved roadway, allowing for a realistic and detailed travel analysis. This approach not only calculates the easiest route but also identifies potential bottlenecks or areas requiring more effort.

Including functional distance in your models is invaluable to tasks like the following:

- » Planning efficient evacuation routes during emergencies
- » Designing hiking trails that balance effort and scenic views
- » Mapping wildlife corridors that avoid high-friction areas like highways or urban centers

REMEMBER

Functional distance calculations expand GIS beyond simple measurement tools, allowing you to model costs realistically in terms of time, money, or even physical and emotional stress.

IN THIS CHAPTER

» Knowing what makes surfaces tick

» Exploring statistical surface types

» Predicting missing values with interpolation

Chapter **12**

Working with Statistical Surfaces

When you hear the word *surface*, you probably picture the ground beneath your feet, or maybe a map's topographic lines. But in GIS, surfaces go beyond the physical topography. They help you map everything from rainfall patterns to population density, turning raw data into visual insights. In this chapter, you explore the different types of statistical surfaces, how to collect and analyze surface data, and how various interpolation methods can help you make sense of your data.

Examining the Character of Statistical Surfaces

In GIS, the term *surface* refers to a dataset that represents a continuous phenomenon across space. Think of a map that shows elevation (a physical surface) alongside a map showing population density (a human surface). Both have the same fundamental data structure but tell entirely different stories: one is about the natural world; the other is about human patterns.

Here's a closer look at these two categories of statistical surfaces:

- » **Physical surfaces:** Think of physical surfaces as nature's blueprint, reflecting the natural world and its processes. Topography (those curvy contour lines on trail maps) is a classic example, but physical surfaces can also represent subsurface phenomena such as groundwater depth, the thickness of rock formations, or even weather-related variables such as temperature, air pressure, and precipitation. Physical surfaces help you visualize and analyze the natural world, both above and below ground.

- » **Human surfaces:** The term *human surfaces* refers to distribution patterns of people, including socioeconomic factors like population demographics (age, income, education), housing costs, taxes, and crime rates. Human surfaces even include mapping attitudes and beliefs like voting trends or economic optimism. But here's the kicker: Human surfaces aren't continuous. Instead, they often occur in distinct locations, which means that you sometimes have to assume that they're spread evenly across space so that you can analyze them effectively.

Every surface, whether physical or human, boils down to the three ingredients depicted in Figure 12-1:

- » **X and Y coordinates:** These coordinates tell you where the data point is located on the map.

- » **Z value:** This number describes the surface by representing the feature's value at that location (like elevation or crime rate). For example, an elevation surface might have Z values of 500 meters in a valley and 1,500 meters on a mountaintop.

- » **Spatial variation:** The surface changes as Z values vary across it, creating high points (peaks), low points (valleys), and flat areas.

REMEMBER

The X and Y coordinates define location, but Z values add another layer of information to your data. In many cases, Z values represent elevation, giving data a true third dimension, but they can also represent any measured value at a location, such as temperature, pollution levels, or even crime rates. Mapping X, Y, and Z values, such as the sample surface shown in Figure 12-1, creates a visual representation of highs, lows, and trends, which makes your data much easier to understand at a glance.

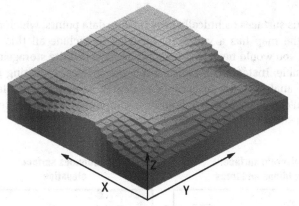

FIGURE 12-1:
X, Y, and Z values of a surface.

kseniyaomega/Adobe Stock Photos

Understanding discrete and continuous surfaces

In GIS, surfaces are either discrete or continuous. In Figure 12-2, the map on the left shows buildings and trees scattered in specific spots (discrete), whereas the image on the right depicts elevation data that flows across the surface (continuous). Here are more details about discrete versus continuous surfaces:

- » **Discrete:** On a discrete surface, data shows up in specific spots but doesn't exist everywhere in between. These surfaces often represent items you can count, such as houses on a block, trees in a park, or crime incidents by address. On a map, discrete surfaces look blocky or pixelated because each data point is separate and distinct. For example, a map of tree cover shows each tree as a distinct point, creating a discrete surface.

- » **Continuous:** With a continuous surface, every part of the surface has a measurable value that blends naturally across the landscape. For example, the image on the right side of Figure 12-2 shows the continuous change of elevation values at Valles Caldera National Preserve, New Mexico, from 1,575 meters in the low areas to 3,525 meters at the top.

REMEMBER

Knowing the difference between discrete and continuous surfaces is important for choosing the right tools in GIS. Discrete surfaces work well for mapping specific, countable features like schools or car accidents. Continuous surfaces, on the other hand, are ideal for visualizing smooth transitions, such as elevation, rainfall patterns, or population density modeled as a heat map.

CHAPTER 12 **Working with Statistical Surfaces** 185

TECHNICAL STUFF Continuous surfaces technically have infinite data points, which means that every spot on the map has a value. If you tried to include all this data in one GIS database, you would need that database to have infinite storage capability, which isn't possible. Instead, GIS tools sample the surface, collecting just enough data points to analyze with an acceptable degree of accuracy while saving storage space. You can discover more about how sampling works in the section "Sampling statistical surfaces," later in this chapter.

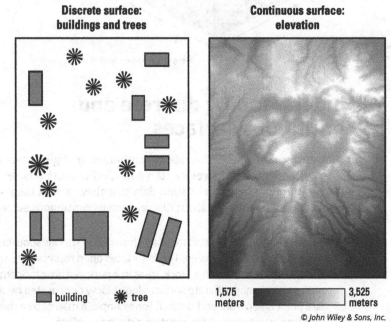

FIGURE 12-2: Discrete versus continuous surfaces.

© John Wiley & Sons, Inc.

Exploring rugged and smooth surfaces

Surfaces don't just vary in their values, such as elevation or temperature, but also in how those values change across space. Some surfaces transition gradually; others change suddenly, as follows:

>> **Rugged surfaces:** Picture a jagged mountain range. Rugged surfaces have abrupt value changes, signaling sharp transitions. For instance, abrupt changes in terrain values signal steep cliffs. Similarly, dramatic swings in soil nutrients can create stark differences in plant growth, with lush vegetation in one spot and stunted crops just a few feet away.

» **Smooth surfaces:** Imagine a gently rolling meadow. Changes in value across smooth surfaces are gradual, making these surfaces easy for people to travel across as well as analyze. For example, gradual changes in elevation create smooth hiking trails; subtle shifts in barometric pressure lead to stable weather patterns.

Knowing whether a surface is rugged or smooth can guide your analysis. Smooth surfaces may require fewer data points, whereas rugged surfaces need more to capture sharp changes accurately.

Climbing steep surfaces

The steepness or slope of a surface tells you how quickly elevation (or any other measured surface value like temperature, air pressure, or pollution levels) changes over distance. The steeper the slope, the greater the impact on movement, construction, or data patterns. For example:

» In the physical world, steep slopes make travel harder, requiring more energy to move uphill. Steep slopes also affect construction, requiring more grading, stabilization, or removal of material to safely access a buried ore deposit.

» In data patterns, steep changes in population density, crime rates, or air pressure highlight areas of rapid change, requiring attention and possible intervention in potential hot spots.

At the beginning of this chapter, I introduce Z values, the values that represent measured attributes, such as elevation or temperature, at different locations. Slope is essentially a measure of how these Z values change across a surface. The steeper the slope, the quicker those values increase or decrease from one location on the surface to another.

Determining slope and orientation

Slope doesn't just tell you how steep a surface is; it also reveals its orientation, or the direction the slope faces. Taken together, these two properties — slope and orientation — give you a complete picture of the surface. Here's the scoop on both of these properties:

» **Slope:** Think of this property as the steepness of a hill. You calculate slope as the change in value of the rise over the run (or horizontal distance). For example, a 20-foot rise over a 100-foot run equals a 20 percent slope. The formula looks like this:

$$Slope = \frac{Rise}{Run}$$

» **Orientation:** This property tells you where the slope points: north, south, east, or west. GIS typically represents orientation in degrees, like this: 0° is north; 90° is east; 180° is south; and 270° is west.

Slope and orientation are important for a variety of environmental and physical processes, influencing everything from climate to landscape changes. For example:

» South-facing slopes in the northern hemisphere get more sunlight and stay warmer, so they're great for vineyards and saguaro cactuses.

» Changes in barometric-pressure slope and orientation influence wind speed and direction.

» Slope and orientation of terrain affect water runoff and erosion, influencing where floods occur and how landscapes change over time.

Working with Surface Data

In GIS, surfaces represent datasets that you can map across space, with values that vary based on location. For example, the temperature highs and lows on a weather map are the Z values, or statistics, that give the surface meaning. Statistical surfaces have the following three properties:

» **They contain Z values distributed across geographic space.** Knowing that your data are distributed across the surface enables you to predict values for areas without sample data.

» **They're measured and recorded at interval or ratio scales.** Interval or ratio scale numbers provide precise statistical numbers, making your analysis more effective than using ordinal measurements like "small," "medium," or "large" would be.

» **You can treat many surfaces as continuous, even if the data isn't everywhere.** Although a continuous surface assumes that data exists at every point (like elevation or temperature), GIS lets you work with discrete datasets, such as population by county, by interpolating values between known points.

These three factors give you the power to predict missing values and group data based on your needs. The following sections show you how to sample and analyze the statistical surface.

Collecting and preparing surface data

Surface data comes from a variety of sources, and how you collect the data depends on the type of analysis you're doing. For physical surfaces, you might collect elevation data using a GPS receiver, drone, or satellite. Tools like LiDAR (Light Detection and Ranging) capture elevation data with incredible detail and are great for topographic maps or flood modeling. (See Chapter 7 for details on collecting data).

For many types of data, such as population, income, or pollution levels, information is often aggregated to regions, like counties or census tracts, rather than measured at every individual point. Having aggregated data is like looking at a map where each county has one value; it's useful but not detailed enough for some analyses.

But here's the challenge for mapping surface data like population density or pollution levels: GIS works best when you can analyze data as points spread across space. Fortunately, there are methods to fill in the gaps. Even if your data is aggregated to polygons (like counties), you can use *interpolation* to estimate values between known points, creating a smoother and more continuous representation of the surface. Instead of looking like a patchwork quilt, your map will display gradual variations, making trends easier to spot and decisions easier to make. Check out the next section for the details on how interpolation works.

Sampling statistical surfaces

With continuous surfaces, the data exist everywhere. But who has unlimited time or resources to work with an infinite amount of data? You can address this problem with sampling. Effective sampling means deciding where to collect data points to get the best picture of your surface layer without overloading your system. Follow these steps to implement effective sampling:

1. **Look at the big picture by studying the surface to find the high and low points in the data.**

   Look for and note any apparent patterns.

2. **Note the smooth and rugged surface types.**

   Smooth surfaces, such as gradual temperature changes, need fewer samples. Rugged surfaces, like jagged cliffs or erratic weather, require more points to capture detail.

3. **Create your sampling plan with a focus on the areas where change happens most often.**

   For example, if you're working with elevation data, take more samples in steep areas but fewer where the terrain is flat.

WARNING

Don't go overboard! Adding too many samples can slow your system and leave you buried in data without providing much improvement in accuracy. Save time and money by sampling effectively and letting the software do the work with fewer, well-placed samples.

TIP

In ArcGIS Pro, tools like `Create Fishnet` or `Generate Random Points` help you create systematic or random sampling grids. For example, if you're analyzing rugged terrain, use `Generate Random Points` to focus sampling on areas with steep slopes identified through a slope analysis layer. In QGIS, use `Grid Vector Layer` or `Random Points in Polygons` plug-ins to achieve similar results. These tools are great for adapting sampling density based on your surface's characteristics. Fire up your GIS tool of choice and give it a try!

Displaying and analyzing Z values

The real fun of working with surfaces begins after you've collected and prepared your data. Analyzing the Z values — numbers like elevation, temperature, or population density — adds depth and dimension, making surfaces feel more real and interactive.

Here are some ways you can use Z values:

» **Drape and overlay:** Imagine that you're studying the ecology in a particular area. You can see how altitude affects plant growth by draping vegetation data over an elevation map. (For information on map overlay techniques, see Chapter 15).

» **Spot the change:** Continuous data lets you track shifts across space, like air or water movement, wildlife migration, or disease spread.

» **Model the future:** Z values help you predict and plan, whether you're mapping watersheds, designing flood control, or deciding where to build infrastructure like bridges or roads.

Studying and analyzing surfaces makes it easy to see trends and relationships that may not be obvious in raw numbers. It's like turning a spreadsheet into a 3D map that brings the story of your data to life.

Challenging the rules of continuous data

I'm going to let you in on a little GIS secret: Sometimes it's okay to break the rules. This is the case with surfaces. They're supposed to be continuous, but what if your data isn't?

Suppose you're mapping population by county. People don't live evenly across a county; they group themselves in cities and are widely dispersed in rural areas. But GIS can still fill in the gaps by interpolating between sample points. Interpolation assumes continuity where it may not exist, creating a modeled surface that helps reveal broad spatial patterns rather than an exact representation of reality. Although it smoothes out gaps in data, you need to remember that interpolation is an estimate, not a perfect reflection of where people actually live.

To understand interpolation pretend that your map is a puzzle. Discrete data, like population by county, are scattered puzzle pieces. Interpolation, which I cover in detail in the next section, fills in the missing pieces, making the surface whole so that you can see the full picture.

When it comes to surfaces, breaking the rules isn't about cheating, but rather about being smart and resourceful. By challenging assumptions, you can uncover patterns and insights that transform your analysis.

Predicting Values with Interpolation

Interpolation is a method used to estimate missing values based on known data points. It works by identifying patterns in the data and using those patterns to predict what's missing. Take a look at these numbers: 10, 20, 30, ??, 50, 60. What's missing? You probably guessed 40 (and you're right, of course) because the numbers follow a predictable pattern. In GIS, interpolation works the same way: It predicts the missing values based on known ones.

Surfaces, like elevation or temperature, often have gaps in the data. To fill these gaps, interpolation methods use patterns in the data to estimate the missing values. Some surfaces change smoothly and steadily, whereas others are more erratic.

The method you choose depends on how your data behaves. This section walks you through the following three categories of interpolation:

- **Linear interpolation:** A simple and useful method for smooth surfaces
- **Non-linear interpolation:** Works well for more realistic and complex data
- **Advanced techniques:** Best for when you need precision and deeper insights

Check out the sidebar later in this chapter for a cheat sheet on the most common interpolation methods and their best uses.

Determining values with linear interpolation

Linear interpolation is the simplest method for estimating missing values. It assumes that your surface changes in a straight line, smoothly and predictably, like a gently sloping hill. Although natural surfaces aren't usually this perfect, the linear interpolation method lays the groundwork for more complex techniques.

Figure 12-3 shows how linear interpolation predicts the elevation (Z value) at a specific point between two known locations (A and B). You start with the facts you know, such as the following:

- Point A is at a 200-foot elevation.
- Point B is at a 400-foot elevation.
- The surface distance between Point A and Point B is 1,000 feet.

Say you want to estimate the elevation at a point 300 feet from Point A (where X marks the spot in Figure 12-3). Follow these steps to accomplish that task:

1. **Identify your points, which in this case are 200 feet at point A and 400 feet at point B.**

   For this example, you're estimating the elevation at 300 feet from Point A.

2. **Find the total elevation change by subtracting the elevation at Point A from Point B; using the example as follows:**

   ```
   400 feet - 200 feet = 200 feet
   ```

3. **Divide the total surface distance into equal chunks.**

   Break the 1,000-foot distance into manageable intervals. For this example, divide it into 10 intervals of 100 feet each:

   ```
   1,000 feet ÷ 10 = 100 feet per interval
   ```

4. **Calculate the elevation change per interval by dividing the total elevation change by the number of intervals:**

   ```
   200 feet ÷ 10 = 20 feet per interval
   ```

5. **Estimate the value at a specific point (300 feet, in this case) using this formula:**

   ```
   Estimated elevation = (number of intervals x elevation
       change per interval) + starting elevation
   ```

   Here's how to break down this formula using the example:

   a. First, figure out the number of intervals this covers:

   ```
   300 feet ÷ 100 feet per interval = 3 intervals
   ```

   b. Calculate the elevation change per interval:

   ```
   3 intervals x 20 feet per interval = 60 feet
   ```

   c. Add the elevation change to Point A's elevation:

   ```
   200 feet elevation at point A + 60 feet = 260 feet
   ```

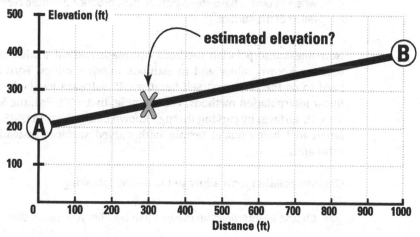

FIGURE 12-3: How linear interpolation works.

© John Wiley & Sons, Inc.

CHAPTER 12 **Working with Statistical Surfaces** 193

REMEMBER

Linear interpolation is great for quick estimates or when you're working with relatively smooth data. It estimates missing values by assuming a straight line between two known points, dividing the space into equal intervals. The smaller the intervals, the more detailed your results will be, but the values always change at a steady rate — no unexpected jumps or dips occur in the values between points.

TIP

When using linear interpolation, break the distance into intervals that make sense for your analysis. Large intervals (like 100 feet) are great for quick estimates, whereas smaller intervals (like 10 feet) give finer details but take more effort.

Using non-linear interpolation

Not all surfaces follow a straight-line path, especially in the natural world, whose features are more complex. Mountains, rivers, and other natural features often change in irregular ways. For example, a riverbed might gradually deepen before dropping off into a steep canyon, or a city's temperature might rise sharply in urban areas but cool suddenly in a park. Non-linear interpolation methods allow you to model variations of the natural world more accurately than a straight-line method by accounting for Earth's curves and irregularities. Instead of treating the space between points as a straight line the way linear interpolation does, non-linear interpolation adjusts for changes in the rate of variation, which means that

» If values increase rapidly in one area and slowly in another, non-linear interpolation adjusts accordingly rather than assuming a steady increase like linear interpolation.

» Non-linear interpolation can recognize that some areas have gradual slopes, whereas others have steep cliffs or dips, producing a more realistic representation of the surface.

Non-linear interpolation methods use mathematical models to detect patterns between known values and to estimate missing values. Fortunately, you don't need to do the math yourself because GIS software has built-in tools for non-linear interpolation methods. For example, in ArcGIS Pro, the Spline tool creates smooth surfaces by curving through known data points. In QGIS, the TIN Interpolation tool helps model terrain with curved surfaces instead of straight-line estimates.

GIS interpolation tools allow you to do the following:

» Choose an interpolation method that best fits your data's characteristics

» Adjust parameters to fine-tune the surface for more accurate predictions

» Create smoother, more natural-looking surfaces instead of rigid, artificial surfaces

To choose the best non-linear interpolation method for your project, start by asking yourself these questions:

» **Is the surface gradual, steep, or a mix of both?**

Some methods, like spline interpolation, work best for gradual changes. Other methods, like kriging, handle more abrupt variations. (See the sidebar "Filling in the blanks: interpolation methods you should know," later in this chapter, for information on common interpolation methods.)

» **Is the surface smooth, rugged, or somewhere in between?**

Choosing the right method depends on whether you need a smooth, flowing surface or one that captures sharp breaks and irregularities.

Non-linear interpolation works great when you know the variations in your surface. With this knowledge, you can match the techniques that provide the most realistic representation of your data.

Estimating with distance-weighted interpolation

One of the fundamental concepts in spatial analysis is Tobler's First Law of Geography, which states that "everything is related, but near things are more related than distant things." The technique known as inverse distance weighting (IDW) applies this principle to predict values by giving more weight to nearby points and less to distant ones when estimating values.

The effects of distance are important for IDW interpolation, as the following examples reveal:

» You're "hangry" (hungry and angry, which is never a good combo) and need a candy bar right now. In this state, you're more likely to dash to the convenience store down the street rather than trek across town to some big grocery store.

» Your favorite band has a concert next week more than 500 miles away, but the following week they're playing in your city. You'll probably catch their show at the closer venue rather than spend the extra time and money to travel 500 miles.

You can always find exceptions to these examples, but generally, the closer you are to your candy bar or your favorite band, the better. Interpolating surface data with IDW works the same way: a point that's five feet away affects the result more than one that's 50 feet away.

IDW is an ideal choice for interpolation if you have scattered data and need to fill gaps without overcomplicating things, or if your surface changes quickly over short distances.

Exploring beyond basic techniques

Sometimes, simple formulas like linear interpolation or IDW aren't enough to capture spatial patterns or produce accurate estimates. Advanced methods like trend surface analysis or kriging are available to help you handle greater complexity, such as when you need to detect how quickly values change or understand the impact of spatial variation and uncertainty.

Here's a quick overview of these two powerful techniques:

» **Trend surface analysis:** Think of this technique as seeing the "forest" instead of the "trees." Trend surface analysis smoothes out the details to enable you to focus on the big picture, like whether a slope is increasing or decreasing overall. It's great for understanding large-scale patterns when you aren't overly concerned with pinpoint accuracy.

» **Kriging:** If IDW is like a trusty old pedal bike, kriging is the high-tech electric bicycle version of spatial interpolation. Kriging takes interpolation to the next level. It focuses on local detail, producing highly precise predictions and even an estimate of uncertainty. Use this technique when you need exact results and confidence in your predictions.

Trend surface analysis might help a city planner identify overall population growth trends, while kriging could pinpoint where water quality issues are most severe. If you're not sure which interpolation method is right for your analysis, take a look at the nearby sidebar for a list of interpolation methods that include these and other methods.

TIP

Start your analysis with the big picture using trend surface analysis to explore general patterns in your data. After you've identified areas of interest, refine your analysis with more advanced techniques for precise predictions and a deeper understanding of details. This two-step approach ensures that you don't miss important trends or local variations.

Try this approach yourself. Interpolation tools are available whether you use ArcGIS Pro or QGIS:

- In ArcGIS Pro, most interpolation tools are found in the Geostatistical Analyst, Spatial Analyst, or 3D Analyst extensions. However, some tools, like Natural Neighbor, are available without an extension.
- In QGIS, look in the Processing Toolbox for the Interpolation toolset. GRASS GIS tools also include interpolation methods, but for full functionality, ensure that GRASS GIS is enabled in the QGIS Plugin Manager.

To find interpolation tools in your GIS software, search by the interpolation method's name like this:

- **In ArcGIS Pro:** Use the Find Tools search.
- **In QGIS:** Use the Processing Toolbox search.

Check your software's documentation or user manual to confirm whether an extension or plug-in is required for a specific tool.

- **Linear interpolation:** This method offers the easiest way to predict values for continuous data, assuming that values change evenly between known points. It's best used for smooth, predictable surfaces or quick estimates.
- **Inverse distance weighting (IDW):** Gives more weight to nearby points. This method is ideal for surfaces with rapid changes of short distances, like soil nutrients or urban temperatures.
- **Trend surface analysis:** This method focuses on the big picture by smoothing out local details to reveal overall patterns. It's great for understanding broad trends like population growth.
- **Kriging:** Your go-to technique when precision matters. Kriging uses spatial relationships to predict values with a measure of uncertainty. It's excellent for environmental monitoring or detailed geostatistical analysis.

FILLING IN THE BLANKS: INTERPOLATION METHODS YOU SHOULD KNOW

To help you determine which interpolation method is best for your analysis, this "cheat sheet" lists the most common techniques and the situations they work best for.

- **Spline interpolation:** Think of this type of interpolation as creating a "rubber sheet" stretched across your known points. Splines create smooth surfaces, making them ideal for elevation mapping or any data requiring smooth transitions.

- **Natural neighbor:** This method looks at nearby data points to create smooth and realistic surfaces without introducing too much complexity. It's a go-to method for environmental data, like rainfall or vegetation.

- **Nearest neighbor:** This is the simplest, but also the crudest, of all interpolation methods. It assigns the value of the closest known point to each unknown location. It calculates quickly and works best for categorical data like soil type or land use.

IN THIS CHAPTER

» Using topography to spot what you can and can't see

» Mapping watersheds and stream basins

» Understanding how topography shapes water flow

» Identifying stream parts

Chapter 13
Exploring Topographic Surfaces

Topography is where Mother Nature unfolds her drama: rolling hills, steep ravines, and everything in between. Topography is where water flows, erosion reshapes the land, and our perspective of the world takes shape, revealing what we can, or can't, see.

But topography isn't just about breathtaking landscapes. Within GIS, understanding topography plays a critical role in solving real-world challenges, from predicting floods and managing water resources to protecting habitats and designing infrastructure.

In this chapter, you get to know three key tools for working with topographic surfaces: viewshed analysis to understand visibility; basin analysis to map water flow; and stream network analysis to study how waterways interact.

Modeling Visibility with Viewsheds

A *viewshed* is a visual representation that shows what is or is not visible from a specific location. For instance, if you're standing outside, your viewshed is everything within your line of sight, although what you can't see may also be important for your analysis. Whether you're planning a scenic overlook, designing cell tower placements, or figuring out how to hide something in plain sight, viewsheds are a go-to tool for analyzing visibility.

When running a viewshed analysis, you analyze visibility in two ways:

- » **Point-to-point:** Is your observation point able to see a specific target? The point-to-point method is perfect for checking whether a newly built fire lookout tower overlooks a critical forested area.
- » **Path-based:** What can you see (or not see) along a route? Hikers use the path-based method to ensure that their trails have great views. Urban designers use this method to avoid visible infrastructure eyesores.

Using viewsheds in the real world

Viewsheds aren't just scenic views; they also help you solve practical problems. Also, they take into account both what you can see and what you can't. Following are some examples of how industries and organizations use them:

- » **Telecommunications:** Viewsheds are used by telecommunications companies to find the best spot for cell towers so that signals reach as many people as possible.
- » **Urban design:** Viewsheds help urban designers hide highways from residential neighborhoods by strategically placing terrain barriers or trees.
- » **Military:** A viewshed can pinpoint hidden areas to safely position troops and equipment.
- » **Real estate:** Real estate developers can use a viewshed to plan a resort with uninterrupted mountain views while keeping unsightly buildings out of sight.
- » **Natural resources:** Viewsheds help the Forest Service, park services, and other agencies locate fire watch towers so that observers can clearly see and monitor large areas for signs of wildfire.

TIP

Deciding whether your project needs a viewshed analysis is simple. Just ask yourself: "Does terrain affect whether I can see certain features?" and "Do I want to see those features?" Then let your GIS tool do its magic! ArcGIS Pro has Spatial Analyst tools like `Viewshed` and `Visibility`, and QGIS has tools like `Viewshed Analysis` and `r.los (Line-of-Sight)`.

Simulating line-of-sight with ray tracing

Imagine that you're in a dark room illuminated only by the flashlight in your hand. The beam of the flashlight reveals what's in front of you while shadows hide what's behind obstacles in your path. That's essentially how a method called *ray tracing* works in GIS. Ray tracing simulates how lines of sight travel across the landscape.

Ray tracing isn't just a GIS tool. Animators and game developers use the same technique to create realistic lighting and shadows in 3D graphics. In GIS, this technique is a big deal for modern applications like the following:

» **Solar analysis:** Architects use ray tracing to model sunlight and shadows to design energy efficient buildings.

» **5G networks:** Telecommunications engineers use ray tracing to find the best spots to place antennas in dense urban areas.

How ray tracing works in GIS

GIS draws lines from your observation point to every potential point in the landscape. If the line reaches the target unobstructed, that spot is visible. If it hits something like a hill or building, it's not.

Figure 13-1 shows ray tracing in action. Areas shaded in gray are hidden, and unshaded areas are visible to the observer.

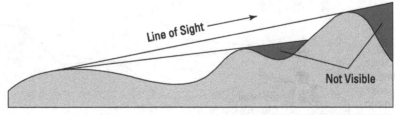

FIGURE 13-1: Simulating line of sight with ray tracing.

© John Wiley & Sons, Inc.

CHAPTER 13 Exploring Topographic Surfaces 201

Viewsheds depend entirely on your perspective. If you move your observation point, your viewshed changes, too, just as it does in real life when you take a step sideways to get a better view of a breathtaking canyon.

Making viewsheds more realistic

In the real world, visibility doesn't pertain strictly to how high you are. Trees, buildings, and other features can block your view, so be sure to take these factors into account by adding their heights to your model:

» **Obstructions:** Include layers with height attributes for trees, buildings, and other barriers. For example, if a tree is 30 feet tall, add 30 feet to its elevation value.

» **Observer height:** If the observer is standing on the ground, add five or six feet for eye level. If observing from a lighthouse (for example), add the lighthouse's height to your location's elevation. Keep in mind that the visibility from a lighthouse is much broader than if you're standing on the ground, so don't forget to account for height!

Figure 13-2 shows how accounting for these factors creates a more accurate viewshed.

 To calculate viewsheds, start with a simple Digital Elevation Model (DEM). Then, if you need more realistic results, add additional details like the observer height and obstructions.

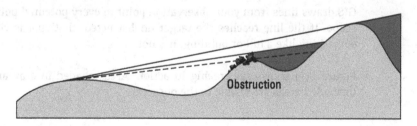
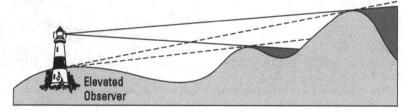

FIGURE 13-2: Make adjustments for observer elevation and obstructions.

© John Wiley & Sons, Inc.

Mapping Watersheds and Basins

Where does the water go when it rains? The answer to this question is critical for things like flood management, erosion control, and environmental planning. You'll often hear terms like "watersheds" and "basins" used to describe areas where water drains to a common outlet, like a stream, river, or lake. Although these two terms are often used interchangeably, there are subtle differences worth noting:

- **Watershed:** This term is commonly used in environmental studies, conservation, and planning. It focuses on water flow and resource management within a defined area.
- **Basin:** This is a broader term often applied to larger or geologically significant drainage systems, particularly in hydrology and geology.

For simplicity, when working in GIS you can think of these two types of draining areas as the same. GIS uses both terms to describe the practical steps for mapping and modeling these systems. No matter the terminology, it's important to understand how water moves and pools within these areas, and GIS is the perfect tool for the job.

Understanding watersheds

A watershed acts like a giant funnel: every drop that falls within its boundaries flows downhill, eventually draining into the same outlet, like a stream or river. The watershed boundaries are defined by the topography of the land, and the way the water moves depends on the type of terrain, as follows:

- **Steep watersheds:** On steep slopes, water rushes downhill rapidly, carving channels and carrying sediment along the way.
- **Gentle watersheds:** In flatter areas, water moves more slowly, pooling in low spots, and creating wetlands.

By using GIS tools like a Digital Elevation Model (DEM), you can analyze how water flows through both steep and gentle watersheds, helping you predict flooding, mitigate erosion, and plan water infrastructure.

REMEMBER

Watersheds are made up of the area upslope from a stream network. Any rain that falls within the watershed boundary can flow overland into the streams below. By mapping watershed boundaries, GIS helps you uncover the dynamics of water flow and its effects on the landscape.

CHAPTER 13 Exploring Topographic Surfaces 203

Working with basins in GIS

To map basins in GIS, you start with elevation data to determine their boundaries. Basins are formed by *ridge lines*, which are the high points in the landscape that separate one drainage from another. Within each basin, water flows downhill toward *pour points*, the lowest points where water drains from one area into another.

Figure 13-3 shows how GIS uses elevation grid data to find ridge lines and model water flow within a basin. Each arrow in the flow direction grid represents the downhill path that water takes from one cell to the next. As water flows downslope, it accumulates in lower cells, forming streams and defining the structure of the basin and its tributaries.

Because basins are closely tied to water flow, mapping them involves the same concepts as those you find in flow-direction analysis (see the upcoming section "Characterizing Water Flow"). GIS tools calculate water-flow paths and accumulation for you, but knowing how this process works helps you visualize basin boundaries and interpret your results more effectively.

TIP

For simpler modeling, start with a "depressionless surface" by filling in minor dips and peaks to create a smooth surface for water to flow over. This ensures smooth water flow and avoids artificial sinks that can complicate your analysis.

TIP

To make your model even more realistic, you can add attributes like precipitation levels or surface absorption rates. For example, areas with high rainfall and low absorption (like paved roads) will accumulate more water, which can help with flood modeling.

FIGURE 13-3: Elevation grid (left) and corresponding flow directions (right).

Elevation

78	72	69	71	58	49
74	67	56	49	46	50
69	53	44	37	38	48
64	58	55	22	31	24
68	61	47	21	16	19
74	53	34	12	11	12

© John Wiley & Sons, Inc.

204 PART 4 Analyzing Geographic Patterns

Characterizing Water Flow

When a raindrop lands on a hill, which way does it roll? Water always takes the path of least resistance, flowing downhill and shaping the landscape as it moves. Understanding how water flows is critical for erosion control studies, flood modeling, and designing structures that can stand up to the forces of nature. GIS tools help you map and model the movement of water by analyzing elevation data to figure out where the water is going and how fast it's moving.

Understanding why flow direction matters

Flow direction is concerned with not only where water moves but also what happens along the way. The direction determines

- **Erosion hot spots:** Where erosion is likely to take place
- **Flood risk:** How quickly the water will move
- **Debris movement:** The potential destructive force of the moving water on important structures in its path

As an example of the importance of flow direction, building in a floodplain may seem like a good idea until you see how water from upstream accelerates downhill, picking up enough force to wash away structures. Modeling flow in GIS helps identify risks like these before they become problems in real life.

Modeling flow direction in GIS

To map water flow, GIS uses a grid of elevation values, typically from a Digital Elevation Model (DEM). Think of this grid as a landscape broken into tiny squares. Each square (grid cell) has an elevation value, and water moves from higher cells to lower ones. Simple enough, right? The following steps, along with Figure 13-4, show how to work with GIS to determine the flow direction:

1. **Start with a target cell.**

   This is the central grid cell (in a 3-x-3 elevation grid) where you want to calculate the direction from which water will flow out.

2. **Compare the neighboring cells.**

   GIS examines the eight surrounding cells and compares their elevation to the central cell's elevation.

CHAPTER 13 **Exploring Topographic Surfaces** 205

3. **Find the steepest descent.**

   The neighboring cell with the largest elevation drop from the central cell is where water will flow. This drop is calculated using the following formula:

   Elevation drop = central cell elevation − neighboring cell elevation

   - The largest positive drop indicates the steepest downhill path.
   - In Figure 13-4, the lowest drop is the Northeast (NE) cell. Water will flow toward this cell.

4. **Assign a direction code.**

   After finding the steepest path, GIS uses the D8 method to assign a direction code to the central cell. The *D8 method* (used by both ArcGIS Pro and QGIS) follows a clockwise system, starting with 1 for east and doubling the value as it moves clockwise around the central cell, as shown on the upper right in Figure 13-5. The Northeast cell has a code of 128, so this becomes the direction code assigned to the central cell. Water flows out of this central cell to the northeast.

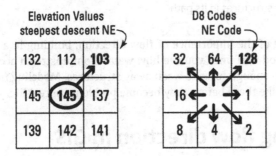

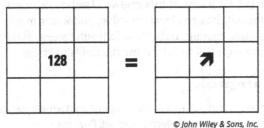

FIGURE 13-4: How GIS calculates flow direction using the D8 method.

© John Wiley & Sons, Inc.

Not every grid cell behaves perfectly. Here are a few quirks you might encounter:

> » **Flat areas:** If all neighboring cells are at the same elevation, water may spread out in all directions. GIS tools often resolve this issue by "filling sinks"

(artificially raising depressions) or assigning flow to one direction to simulate movement.

- » **Local depressions:** If all neighbors are higher, the water can pool, creating a "sink" in your model.
- » **Multiple paths:** If several cells have the same steepest drop, the flow may split.

Different GIS software handles these challenges in unique ways. Some expand the search area to find the best drain point; others apply coding schemes to estimate flow direction.

Determining flow speed in GIS

Sure, it's great to know where water flows, but what about how fast it's rushing across the landscape? The speed can mean the difference between a gentle stream or a destructive flood.

Here's a simplified formula to estimate flow speed in GIS:

Flow Speed = Change in Elevation ÷ (Distance ×100)

This formula provides a rough idea of how elevation differences influence movement, though real-world flow speed also depends on factors like terrain roughness and water volume.

On a 3-x-3 grid, the distance between cells depends on their position:

- » **Adjacent cells:** Distance = 1 (for example, north, south, east, west neighbors)
- » **Diagonal cells:** Distance = 1.414 (a longer path across the grid)

REMEMBER

The Flow Speed formula gives you a general idea of how fast water will move. If you're modeling a steep hillside, you'll see higher flow speeds, which can indicate areas prone to erosion. On flatter terrain, the flow slows, often leading to pooling or sediment buildup.

Modeling flow direction and speed isn't just an academic exercise. It has real-world uses like these:

- » **Flood prediction:** You can map how quickly water will move across a landscape during heavy rainfall.

CHAPTER 13 **Exploring Topographic Surfaces** 207

- » **Soil conservation:** You can identify areas at risk of erosion and plan countermeasures, like planting vegetation.

- » **Infrastructure design:** You can ensure that roads, bridges, and buildings can withstand fast-moving water.

By understanding how water flows, you can plan better, design smarter, and protect resources more effectively.

TIP

Start with a DEM to model basic flow direction. If you need more accuracy, add extra data like precipitation, soil absorption rates, or barriers like levees to refine your results.

Defining Streams

Streams are nature's highways, carving through landscapes, collecting rainwater, and branching into tributaries that create complex networks. These networks don't just move water; they shape ecosystems, control water flow, and even influence the land around them.

Whether you're studying erosion, managing water resources, or protecting habitats, mapping streams in GIS helps you analyze these networks and their connections.

Understanding stream networks in watersheds

Streams are part of larger systems called watersheds. As mentioned in "Understanding watersheds," earlier in the chapter, a watershed is like a giant funnel, encompassing an area where all water in it drains to a common point, such as a stream or river. Take a look at Figure 13-5 showing the key parts of a stream network:

- » **Trunk stream:** The main channel into which smaller streams flow
- » **Tributary streams:** Smaller streams that flow into and feed the trunk stream
- » **Sub-basins:** Smaller sections of a watershed that drain into a larger stream, river, or reservoir within the same watershed
- » **Pour points:** The lowest points where water converges and exits into the trunk stream or leaves the watershed

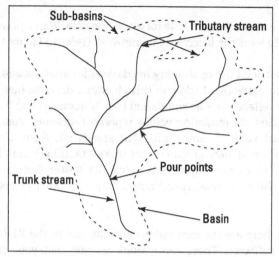

FIGURE 13-5: The parts of a stream watershed showing how water connects across the landscape.

Locating streams in GIS

When working with vector data, finding streams is as simple as selecting lines labeled as streams. But with raster data, you can do so much more than locate streams. You can model their flow, identify tributaries, and analyze how water accumulates in a basin.

Streams form naturally in areas where water collects and flows downhill, often following low spots in elevation. To locate these streams in GIS, you use a Digital Elevation Model (DEM) and follow these basics steps:

1. **Define the basin edges.**

   These are the highest points that form the boundaries of a drainage basin (think of them as the "rims" of a funnel, where all water within an area will eventually flow to a single stream, river, or reservoir).

2. **Identify the pour points.**

   Pour points are the lowest points in each sub-basin where water drains and joins a stream. These junctions mark where smaller tributaries meet.

3. **Trace the streams.**

   Work upward from the pour points, identifying paths of water flow based on elevation.

CHAPTER 13 **Exploring Topographic Surfaces** 209

REMEMBER

A pour point acts like a drain hole for a specific area. Water flows downhill into it and eventually joins the larger stream network (refer to Figure 13-5).

TIP

Here's a little trick I use to simplify my dataset for analysis and make it easier to isolate stream networks. Earlier in this chapter I describe how GIS assigns grid cell values to reflect water accumulation (see "Characterizing Water Flow"). Cells with the highest accumulation values represent streams. You can assign these cells a distinct value — something that stands out from other values in the dataset — and set all non-stream values to NODATA. (Check out Chapter 16 for how to assign values to raster grid cells using the Raster Calculator). This trick is especially useful when overlaying stream data with other layers, like land use or precipitation.

For example, here are the expressions you can use in the Raster Calculator tool of your GIS software. These expressions assume that your raster filename is flowAccum.tif and assign a value of 1 to streams (identified here as cells with a flow accumulation greater than 1,000). All other cells are automatically set to NODATA:

» **ArcGIS Pro:** In ArcGIS Pro, the Con( ) function assigns a value (1 in this case) to cells that meet the condition (flow accumulation > 1,000), while setting all other cell values to NODATA:

```
Con("flowAccum.tif" > 1000, 1)
```

» **QGIS:** In contrast to ArcGIS Pro, QGIS doesn't require a special function; instead, it automatically assigns NODATA to cells that don't meet the condition specified in the expression:

```
("flowAccum.tif" > 1000) * 1
```

These expressions assume that streams have high flow accumulation values, but you can adjust the threshold (1,000 in this example) depending on your dataset and study area.

Identifying methods that work for you

Streams are rarely isolated. They branch into tributaries and merge to form larger streams, creating stream networks. To study these networks, GIS uses a method called *stream order*, which ranks streams based on their connections. Stream order breaks down like this:

» **First-order streams:** These streams are the smallest ones with no tributaries. First-order streams form directly from overland flow, springs, or snowmelt.

For instance, many small, unnamed streams in mountainous areas or headwater regions, like the source streams of the Colorado River in the Rocky Mountains, are first-order streams.

» **Higher-order streams:** These streams form when streams of the same order combine. Two first-order streams merge to form a second-order stream, two second-order streams merge to form a third-order stream, and so on.

For example, you might find a second-order stream where two first-order mountain streams merge in the Sierra Nevada or Rocky Mountains. The Amazon River is an example of a very high-order stream, reaching 12th order — the highest in the world.

The Colorado River is also a high-order stream, reaching 8th order by the time it enters the Grand Canyon, and increasing further as it collects more tributaries. Historically, the Colorado River flowed into the Gulf of California (also called the Sea of Cortez), but because of extensive water diversions, it now rarely reaches the sea.

You can calculate stream order in several ways, but GIS software typically uses one of two ways to calculate it, as follows (and see Figure 13-6):

» **Strahler method:** Each stream level increases only when two streams of the same order meet. For example, two first-order streams combine to form a second-order stream. This method is great for tracking flow strength.

When two different stream orders come together, GIS assigns the higher of the two to the next resulting tributary. For instance, when a first-order and second-order stream come together, they become a second-order stream.

» **Shreve method:** In this method, each stream contributes to the overall order, making it a better fit for ecological studies that consider every tributary.

When stream orders come together, the values are added together, resulting in what's known as "stream magnitudes." For example, when a first-order stream and second-order stream meet, the resulting stream is a third-order stream.

REMEMBER

If you're modeling sediment transport, use the Strahler method to identify high-flow streams. For ecological studies, the Shreve method gives you a better picture of overall stream connections.

TIP

You can use stream order to define buffer zones around streams. Higher-order streams often need larger buffers to protect their flow and surrounding habitats.

CHAPTER 13 **Exploring Topographic Surfaces** 211

FIGURE 13-6:
The Strahler and Shreve methods analyze stream order differently.

Strahler   Shreve

© John Wiley & Sons, Inc.

Most GIS software, like ArcGIS Pro and QGIS, include built-in tools for locating streams, defining watersheds, and calculating stream order. Here are a few tips to consider before you dive into the tools:

» **Check your DEM resolution.** Higher resolution provides more detailed stream networks.

» **Review your software documentation.** Look for tips on handling flat areas, sinks, and flow anomalies.

Use stream order analysis to plan buffer zones, identify key tributaries, or estimate water flow through a network. Stream order analysis is an essential tool for anyone working with hydrology, conservation, or land use planning.

> **IN THIS CHAPTER**
> » Understanding how connected networks work
> » Tackling traffic jams and resistance
> » Mastering one-way streets and tricky turns
> » Planning routes, from shortest to scenic

Chapter 14
Working with Networks

Networks are all around us in the form of roads, railroads, rivers, and even power grids. Of course, networks aren't just lines on a map; they are the pathways that keep our world moving. Whether you're planning the quickest delivery route, designing efficient power grids, or mapping scenic hiking trails, understanding networks is essential for making smarter, data-driven decisions.

In this chapter, you find out how GIS helps you analyze and optimize networks. From calculating connectivity to navigating one-way streets, you discover the tools that make GIS a powerhouse for modeling and measuring movement along networks.

Measuring Connectivity

How well is a network connected? That's the key question in measuring network connectivity. *Connectivity* tells you how well a network's components — links and nodes — are connected. GIS analysts use connectivity to evaluate a road network's efficiency or the reach of a power grid, for example. One common way to measure connectivity is with the *gamma index*, which is the ratio of the number of actual connections in a network to the maximum possible connections. The index ranges from a value of 0 (no connections) to 1 (completely connected).

For example, a road network in a rural area with only a few direct routes between towns might have a gamma index of 0.2, meaning that it has only 20 percent of the possible connections. On the other hand, a well-connected suburban street grid might have a gamma index of 0.5, whereas a dense urban core with multiple intersections and alternate routes might reach 0.8. In real life, a road network rarely, if ever, reaches 1 because of geographic obstacles, land-use restrictions, and the inefficiency of excessive intersections.

Recognizing why connectivity matters

Understanding connectivity goes beyond numbers. It involves grasping how well a network supports movement and flow. A highly connected network makes getting around easier. Meanwhile, anyone who's driven through a city knows that a not-so-well-connected network forces you to go out of your way through a series of long detours.

Any road map of the United States shows you a connectivity pattern. Coastal and other highly populated regions tend to have tightly connected road systems, with many towns linked by multiple routes. The gamma index for these areas is very close to 1. In contrast, the Great Plains region (western Minnesota to Montana), with its sparse connections, has a much lower gamma index, requiring longer routes to get from place to place.

Connectivity encompasses more than roads, though. It also plays a role in the following fields:

- » **Urban development:** You can evaluate how road networks evolve as towns grow and new connections are built.
- » **Environmental planning:** You can track how wildlife moves through corridors or how edge species — those living on the edges of different habitats — expand into new environments.
- » **Infrastructure resilience:** You can ensure that utility grid networks can continue to serve homes and businesses, even if parts of the system fail.

Calculating the gamma index

A network consists of *links* (connection lines, also referred to as edges) and *nodes* (the spots where the links connect), with the gamma index describing how well the network is connected. Calculating the gamma index is straightforward: You count the actual links between nodes and divide that number of possible links. Here's the general formula:

$$\text{Gamma Index} = \frac{\text{Actual Links}}{\text{Maximum Possible Links}}$$

If you're wondering, "How do you know the maximum possible links?" a formula exists for that, too. For a typical road network where streets meet at intersections, it works like this:

$$\text{Maximum Possible Links} = 3(N-2)$$

where N is the number of nodes (intersections or junctions).

The gamma formula can reveal a lot about a network's efficiency. In Figure 14-1, dots represent towns (nodes), and connecting lines represent roads (links). From left to right, the three networks show how the number of connections increases as more roads are built. (If you want a deeper dive into calculating the gamma index, check out the nearby sidebar "Planar networks and the gamma index").

The gamma index is useful for

- » **Characterizing the loss of shelterbelt networks:** Rows of planted trees called *shelterbelt networks* protect crops from the wind. Using the gamma index, you can identify regions where shelterbelts are sparse and prioritize replanting efforts.

- » **Evaluating road networks over time:** As Figure 14-1 illustrates, you can compare connectivity across different years and track how urbanization increases linkages between towns.

- » **Analyzing utility networks:** A highly connected utility grid ensures that electricity, water, or gas flows smoothly. Analyzing the network connectivity to determine where to place substations ensures that if part of the system fails, other connected pieces can continue providing service to customers.

REMEMBER

Connectivity analysis isn't just an academic exercise; it's a tool for solving real-world problems. For example, urban planners might use the gamma index to identify underserved neighborhoods, while utility companies use it to make sure their systems remain operational. Whatever the application, connectivity is a useful tool for understanding and improving the networks that keep the world running.

TIP

To calculate the gamma index with ArcGIS Pro or QGIS, you'll need to calculate it directly using an attribute table or analysis tools. For example, to convert intersections to points in ArcGIS Pro, you use the Feature to Point tool, and in QGIS, you use Extract Nodes. Regardless of which software you use, though, be sure to enable network topology (see Chapter 5 for details on using topology in GIS) and prepare your network data with the attributes needed for identifying nodes and links.

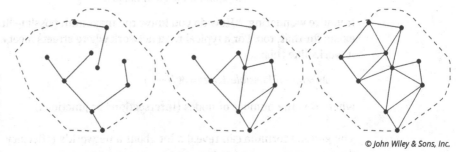

FIGURE 14-1: The gamma index increases while connectivity increases.

© John Wiley & Sons, Inc.

PLANAR NETWORKS AND THE GAMMA INDEX

You may have heard the term "planar network," but what does it actually mean? A *planar network* is a type of network in which no two links cross each other unless they meet at a node (a connection point). Imagine a map of city streets - in a planar network, all intersections are *clean*, meaning that streets either meet at nodes (like crossroads) or remain separate (like parallel roads). In GIS, modeling planar networks relies on topology (see Chapter 5) to ensure that connectivity rules, such as road intersections, are enforced for accurate analysis.

In contrast, a *non-planar network* in GIS is a network whose lines or features cross each other without forming a defined intersection. On a 2D map, they may appear to pass through one another, but in reality, they exist at different elevations or layers, like an underground utility network whose pipes cross at different depths, or an airline network whose flight paths overlap at different altitudes in the sky. Because non-planar networks don't rely on physical intersections, they can be more complex to model in GIS, often requiring elevation data, 3D modeling, or specialized network analysis tools to accurately represent connectivity.

Here's a quick look at the difference between a planar and a non-planar network:

- **Planar network:** A neighborhood street network whose roads all intersect at designated intersections (nodes).
- **Non-planar network:** A subway or airline network where different routes can pass over or under each other without directly connecting.

A planar network has physical constraints. So you calculate its connectivity using the basic gamma index formula:

$$\text{Gamma Index} = \frac{L}{3(N-2)}$$

Here's what those symbols mean:

- *L*: The number of actual links (the connections between nodes in your network)
- *N*: The number of nodes (the points where the links meet)

For non-planar networks, in which every node can theoretically link to every other node, you need to use a different gamma index formula to determine connectivity. The following formula accounts for the fact that no physical constraints prevent every node from being directly connected:

$$\text{Gamma Index} = \frac{L}{\left(\frac{N(N-1)}{2}\right)}$$

Recognizing whether a network is planar or non-planar helps in evaluating network efficiency and limitations. Although planar networks are built around physical constraints, non-planar networks provide greater flexibility by allowing links to cross without direct connections. Understanding these differences helps in designing and analyzing networks for best performance and accessibility.

Working with Impedance Values

Networks are corridors that allow things to move through them: cars on roads, water in pipes, and even emails through the internet. But not all corridors are created equal. Some provide clean and clear pathways, whereas others are filled with obstacles that create resistance. This resistance to movement along networks is called *impedance*.

REMEMBER

Impedance comes in a variety of forms. It might be a bumpy dirt road slowing down your vehicle, a narrow water pipeline restricting flow, or a congested intersection making traffic crawl along at a snail's pace. The most common use of network impedance is in modeling transportation networks, but it can occur in any type of network. Whatever the network, impedance is the factor that decides how fast, or even whether, something moves from point A to point B.

CHAPTER 14 **Working with Networks** 217

Understanding why traffic is good or bad

Ever been stuck in traffic and wondered why things are so backed up? Or wondered why your delivery always runs late? Look no further than impedance. Here are some of the common culprits slowing down traffic networks:

» **Speed limits:** Posted speed limits set the upper (and sometimes lower) bounds for how fast you can go.

» **Traffic density:** Anyone who has ever sat in rush-hour traffic knows that heavy traffic can make even the fastest highway feel like a parking lot.

» **Accidents and disruptions:** Lane closures and detours can throw a monkey wrench into your travel plans.

» **Road conditions:** Construction zones, potholes, or rough gravel roads slow you down.

» **Checkpoints:** Border-crossing checkpoints, sobriety checkpoints, weigh stations, and other regulatory facilities cause delays.

» **Weather impacts:** Rain, snow, fog, and dust storms can wreak havoc on travel times.

While these issues can frustrate drivers, GIS provides tools to analyze and manage them. By modeling real-time conditions, GIS can reroute traffic around construction zones, send alerts about weather-related delays, and help planners optimize road networks. But to make traffic modeling work, you need to configure your network dataset properly. I show you how to do just that in the remainder of this chapter.

Modeling impedance in traffic networks

Understanding and modeling impedance improves your ability to know how long it will take for a fire engine to get to an emergency or how quickly someone's groceries will be delivered. GIS makes this possible with the ability to add real-world conditions directly into a network dataset.

REMEMBER

It's important to note here that your analysis is only as good as your data. You'll need real-world data to account for the impedances along your network. Gathering this data may take time, but it's a critical step in any GIS project.

For impedance data along transportation networks, you have several data options. For example, have you ever noticed those black cables stretched across roads? Those are traffic counters. Transportation analysts use them to measure traffic flow, which is an essential data point for modeling impedance. Other common data inputs include speed limit zones, stop signs, and traffic signals.

TIP

Many governments provide traffic flow and transportation network–related data via open-source data portals. For a list of transportation-related data sources to get you started, see the last section in this chapter, "Working with networks in GIS."

Exploring key impedance layer settings

When setting up GIS for traffic modeling, you can (and should) include various attributes in your impedance layer. The process for defining impedance attributes varies depending on the GIS software you're using:

» In ArcGIS Pro, working with network datasets requires the Network Analyst extension, which provides built-in tools for modeling impedance attributes and network settings.

» In QGIS, the Network Analysis toolbox (part of the core QGIS install) offers basic network modeling functionality but has limited support for advanced impedance settings. However, you can extend QGIS capabilities by installing plug-ins like QNEAT3 or AequilibraE, which add more functionality for working with network datasets.

REMEMBER

Terminology often varies among software applications. In ArcGIS Pro, "cost" is often used interchangeably with "impedance." QGIS mostly uses "cost" to refer to impedance, but the terminology can vary among plug-ins. So always check the documentation of the specific tool or plug-in you're using to understand the exact terminology used and functionality supported.

To help with configuring impedance settings in network datasets, here's a handy list of items to consider and how to implement them in GIS:

» **Impedance attribute:** Defines the cost of travel, such as how long it typically takes to travel a certain distance.

- **ArcGIS Pro:** You set impedance as a cost attribute in the network dataset properties. You can define travel time, distance, or other costs using numerical fields.

- **QGIS:** The cost attribute is assigned when you set up network analysis tools, such as Shortest Path (in the QGIS Processing Toolbox) or when using plug-ins like QNEAT3 or AequilibraE.

» **Default cutoff value:** A configuration setting that defines the maximum allowable travel distance or time for a route, preventing routes from exceeding a predefined threshold.

- **ArcGIS Pro:** You set cutoff values in the properties of network analysis layers (such as a Service Area, Origin-Destination Cost Matrix) to limit travel distance or time.

CHAPTER 14 **Working with Networks** 219

- **QGIS:** The QNEAT3 plug-in includes cutoff values as a setting in its routing tools, allowing you to set a maximum travel distance or time when calculating routes.

>> **Accumulation information:** Attributes that build up over a route, like toll fees, fuel costs, or the number of bus passengers.

- **ArcGIS Pro:** You set accumulation attributes in the network dataset properties by specifying which cost attributes to sum along a route.
- **QGIS:** Use the AequilibraE plug-in to specify which travel costs to accumulate during network analysis.

>> **Restrictions:** Rules about which types of traffic can use certain roads, such as trucks with hazardous cargo being restricted from residential streets.

- **ArcGIS Pro:** You define restrictions as Boolean values (for example, Hazardous Cargo = Yes or No) in the network dataset properties and used in routing analysis.
- **QGIS:** The QNEAT3 plug-in allows for filtering roads based on specific attribute values, but the AequilibraE plug-in offers more advanced restriction handling for network modeling.

>> **Hierarchy:** A setting that prioritizes certain road types over others, such as preferring highways over local roads.

- **ArcGIS Pro:** You set hierarchy in the network dataset properties, allowing analysis to favor major roads over minor ones.
- **QGIS:** QGIS does not natively support hierarchical routing, but you can simulate hierarchy-based routing in the QNEAT3 or AequilibraE plug-ins by assigning higher impedance values to local roads.

After you've configured impedance values in your network dataset, you can apply them to your analysis. One of the most powerful ways to do that is with an *Origin-Destination* (OD) *matrix*, using your impedance settings, like travel time, restrictions, and road hierarchies, to calculate the most efficient routes between multiple locations.

Entering the matrix

The OD matrix just mentioned takes the origins, destinations, and barriers in a network and calculates the most efficient routes based on impedance data. It knows where a trip starts (origin), where it ends (destination), and what gets in the way (meaning barriers, such as road closures or traffic restrictions). By factoring in real-world constraints, OD matrices help optimize movement across a network.

Use OD matrices in your network analyses to help

- » Avoid congested streets during rush hour by factoring in real-time or historical traffic patterns
- » Predict travel times by accounting for toll roads or speed limits
- » Optimize distribution networks by simulating the flow of water through pipelines or electricity through grids
- » Improve transportation planning by calculating the most efficient bus or train routes between high-demand locations

GIS software like ArcGIS Pro and QGIS generate OD matrices that analysts use to support transportation, logistics, and infrastructure planning. In ArcGIS Pro, OD matrices are created using the Network Analyst extension, whereas QGIS users can generate them with plug-ins like QNEAT3.

REMEMBER

The more accurate your impedance data, the better your GIS will reflect the real world, and the more useful your results will be. Consider consulting a subject-matter expert to determine which values to include in the OD matrix. They can help you choose appropriate input values such as speed limits for roads, flow capacities in a pipeline network, or electrical resistance in a utility network.

Navigating One-Way Paths

Every path is unique, and some let you travel in only one direction. These *unidirectional paths*, like one-way streets or flowing rivers, can make navigation tricky but they're critical components for many networks. GIS helps you account for these types of paths in your network, ensuring that your routes follow the rules and you don't end up going the wrong way down a one-way street.

Understanding one-way systems

In some cases, movement naturally flows in one direction. Think of a fast-moving river pushing a raft downstream or a parking lot exit with those nasty tire-shredding spikes to keep vehicles from entering (or backing up). But the most common is the one-way street.

One-way streets are everywhere, and often for good reasons. They manage traffic flow, reduce congestion, and improve safety. However, they also bring their fair share of headaches, such as when your GPS sends you in circles because it doesn't know you can't turn left. That's where properly configured GIS data comes in, helping to keep your trip progressing smoothly.

Incorporating one-way paths in network models

GIS software like ArcGIS Pro or QGIS doesn't automatically know which paths are one way; instead, they rely on you to provide the necessary direction information. Network datasets typically include a layer with directional attributes, such as one-way street flags or other directional rules, which are added during the network dataset setup. Here's how to incorporate on-way paths in your network:

1. **Define the direction rules.**

   Defining the rules involves including directional attributes in your data, such as fields indicating whether a street allows travel in both directions, only one direction, or is closed to a specific vehicle type.

2. **Set up the network dataset.**

   When you create a network dataset in either ArcGIS Pro or QGIS, configure the directional attributes as part of the rules that govern movement. Configuration might involve linking the directional data to your impedance layer or adding it as a stand-alone set of restrictions.

3. **Apply restrictions.**

   With your attributes in place, GIS tools will enforce the rules when routing and modeling networks. For example:

   - GIS will treat a one-way restriction like a virtual barricade, blocking traffic from moving in the wrong direction or not allowing streams to flow uphill.
   - Routing tools will automatically reroute paths to comply with the rules you have established.

Say that you're mapping routes for a fleet of delivery trucks. Including one-way route data into your GIS ensures that drivers avoid making illegal turns or traveling the wrong way down one-way streets. Your data input and the software's processing power work together to create efficient, real-world routes.

Accurately mapping one-way paths not only keeps your routes compliant with traffic laws but also avoids wasting time and fuel. By including directional attributes in your dataset and configuring your network properly, you ensure that your GIS can handle even the trickiest navigation scenarios, whether for city streets, rivers, or pipelines.

Defining Circuitry

Networks are characterized by their complexity. The gamma index, described earlier in this chapter, provides a way to measure this complexity. Another characteristic of networks is *circuitry*, which influences both network connectivity and robustness in traffic modeling. Circuitry refers to the presence of closed loops in a network, allowing alternative travel paths. Think of a roundabout, a bypass, or even a river flowing around an island. Each of these circuits offers an alternative route within the network.

Figure 14-2 shows the difference between an open network and a closed circuit:

» An open network has no fully connected loops, meaning that travel is limited to a single path with no alternative routes.

» A closed circuit includes at least one loop, providing redundancy and improving flow efficiency.

Closed loops provide flexibility in a network and play a crucial role in improving flow and resilience. The following sections show how closed loops work.

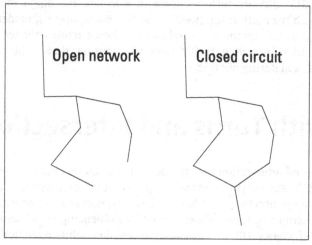

FIGURE 14-2: An open network versus a closed circuit.

© John Wiley & Sons, Inc.

Understanding how circuits improve networks

Circuits are ubiquitous. If you're modeling streams, you may come across an area where a stream splits around an island, moving in two different directions. In many cities, roundabouts create small circuits that keep traffic moving. In the United States, highway bypasses and loop systems do the same thing but on a larger scale, allowing vehicles to avoid slow-moving, inner-city streets.

In GIS, circuits increase your route options. For example, even if a looped path is longer, it may save time by reducing traffic-related delays. These closed loops also improve connectivity and redundancy, making networks more efficient and resilient.

Measuring and modeling circuits

To measure circuitry, GIS uses the *alpha index*, a ratio that compares the number of actual circuits to the maximum possible circuits in a network. Whereas the gamma index (explained in "Measuring Connectivity," earlier in this chapter) measures overall connectivity, the alpha index focuses specifically on loops. Here are some examples of using the alpha index:

- A sparse network with no loops (circuits) has an alpha index near 0.
- A dense network with multiple circuits approaches an alpha index of 1.

Most GIS software includes tools to calculate the alpha index automatically. Although the math is fun (well, if you like math, that is), modern GIS users typically focus less on the math and more on how circuits enhance real-world situations, like improving traffic flow or ensuring that utility networks remain operational during outages.

Working with Turns and Intersections

Turns and intersections are the decision points in a network: Will you turn right or left? These are the points where paths meet, and movement along the network can change direction. Whether the decision point is a T intersection, a roundabout, or a sharp turn, it can impact traffic by influencing travel time, congestion, and route efficiency. GIS helps you model decision points realistically so that you can plan optimal routes.

Recognizing the importance of turns and intersections

When you plan a trip, you need to know where you're starting, where you're going, what the traffic will probably be like, which roads or streets to take, and where you turn from street to street. Turns and intersections affect all these considerations — from your travel time to your safe arrival. Some intersections, such as four-way stops, slow you down. Others, such as freeway on-ramps, speed you up but still require careful navigation as you merge into traffic. And don't forget one-way streets, which force you to plan your turns carefully and strategically.

When planning routes, you have many factors to consider. For example, planning a trip that involves mostly making right-hand turns can save you from having to wait for cross-traffic to clear. If you're determining emergency routes for first responders, knowing the layout of intersections along potential routes will help to improve response times. Modern GIS tools take these factors into account automatically, but the key to successful traffic network modeling is good data.

WARNING

My emphasis on good data may seem repetitive, but good data is the driving force behind GIS, so I can't overstate the need for it. You've probably heard the phrase "garbage in, garbage out" before. Bad data (garbage in) will lead only to bad results (garbage out). Good data, when properly configured, will give you good results.

Encoding and using turns and intersections

When it comes to modeling movement, network datasets have three key components:

- **Edges:** The line work (or links) in your network, such as roads, pipes, or railways
- **Junctions:** The points (or nodes) where edges connect, like a road intersection or pipeline joint
- **Turns:** The movement between edges at a junction, such as turning left, making a U-turn, or simply going straight

The number of possible turns at any junction depends on the number of connecting edges. GIS networking tools use this concept to determine how vehicles or water, for example, move through the network. Figure 14-3 shows a T intersection composed of three connecting edges. Having three connecting edges allows for nine possible movements (three edges squared = 3^2 = 9), as follows:

- **Three U-turns:** Each edge provides an opportunity to do a U-turn if you allow it.
  - **Two right turns and two left turns:** The link coming down from the top allows you to make right turns along the inside lane and left turns to the outside lane.

>> **Two straight-through movements:** You can continue along either of the bottom edges without turning.

Not all of these movements may be practical or legal, though, which is where turn rules come into play. GIS has tools to define what happens at junctions to make sure that your network reflects real-world conditions, such as the following:

>> **Turn directions:** Are left turns allowed? What about U-turns? You can't set up turns that include impossible movements or violate traffic laws.

>> **Turn impedance:** How much time does a turn add to the route? For example, left turns across heavy traffic take longer than right turns.

You store these rules in the network dataset, often as part of an attribute table. Some GIS tools even let you assign impedances to certain turns, like sharp angles or intersections with stoplights, to improve route calculations. By adding this information to your network dataset, your network routing model becomes more realistic.

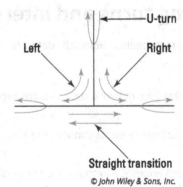

FIGURE 14-3: The possible turns in a T intersection.

© John Wiley & Sons, Inc.

Directing Traffic and Exploiting Networks

If you've ever used Google Maps or Waze to get directions to where you're going, you've seen routing in action. These modern navigation aids are based on GIS technology. But working with networks directly in GIS takes routing to a whole new level, empowering you to analyze, optimize, and manage networks in ways that solve practical problems. Unlike Google Maps, GIS lets you build customized networks tailored to your specific needs, whether you're planning water pipelines or optimizing bike trails.

Here are some of the networking options available in GIS:

- » Finding the best route (the shortest, fastest, or even the most scenic)
- » Finding the closest geographic feature
- » Finding service areas, such as what parts of town are served by a single fire station

Finding the shortest path

You use a method called *shortest-path analysis* to minimize distance. Shortest-path analysis is the go-to method for projects whose cost or effort increases with distance. Unlike consumer apps, GIS allows you to calculate the shortest path for any kind of network, not just roads. Whether it's pipelines or utility grids, GIS can handle it.

Suppose you're working with a telecommunications company. As a GIS analyst, your task is to find the shortest routes for laying fiber-optic cables. The shorter the path, the less equipment and fewer resources needed for the project. Using GIS, you analyze the length of every segment in your network, compare different route options, and identify the shortest and most direct route from point A to point B. Your analysis will not only minimize costs but also help to make the installation process more efficient overall. Plus, with all the cost savings, you might even get nominated for Employee of the Year!

REMEMBER

The shortest-path method works best when distance is the only priority; you don't have to account for impedances like speed limits or traffic congestion.

Finding the fastest path

The shortest path isn't always the quickest. Sometimes congestion, low speed limits, or other impedances can make the shortest path take more time, as anyone who's ever been stuck in traffic can attest to. Your in-car GPS navigation typically uses a method called *fastest-path analysis* by default, connecting to services that provide real-time traffic conditions, speed limits, and even road construction detours. Fastest-path analysis uses impedance values and is ideal for situations for which time is of the essence.

Delivery services are a perfect example. They typically use fastest-path analysis to optimize routes, but it isn't just for pizza and packages. Emergency responders use this method to reach incidents faster, potentially saving lives. Here's how to perform fastest-path analysis in GIS:

1. **Add impedance values to your GIS network dataset.**

   Inputs like travel speed, traffic delays, or stoplight timing allow GIS to determine the best results.

2. **Use GIS to calculate the route with the lowest total travel time, factoring in these real-world conditions.**

   In ArcGIS Pro, use Find Best Route with the Network Analyst extension; in QGIS, use the QNEAT3 plug-in for route calculations.

TIP

Many GIS tools allow you to model time-of-day variations, such as rush-hour traffic or seasonal road closures, for more accurate results.

Finding the nicest path

Sometimes the journey is more important than the destination. With *nicest-path analysis,* you can prioritize routes based on custom attributes, like scenic beauty, safety, or environmental impact. Many national parks now use GIS to create interactive maps that highlight scenic routes and include real-time road conditions. These maps are great both for planning a "nicest-path journey" and avoiding congestion caused by that one visitor who stops to pet a bison, creating a traffic jam and one angry bison.

Angry bison aside, you can follow these steps to create a nicest-path analysis in GIS:

1. **Add custom fields to your network dataset.**

   Define what makes a route "nice." Common fields include:

   - ScenicScore: Rates how scenic a road is
   - SafetyLevel: Rates how safe a road is
   - EcoImpact: Represents environmental impact

   In ArcGIS Pro (Network Analyst extension), open the Network Dataset Properties and add new cost attributes in the Travel Attributes section.

   In QGIS, use the Field Calculator in the Processing Toolbox to create new attribute fields in your road network layer.

2. **Assign values to each segment based on your priorities.**

    Each road segment in your network layer needs a numeric score for the newly added fields. Use real-world data, expert input, or estimates to determine the values. For example,

    - ScenicScore: Assigns higher values to roads with scenic views
    - SafetyLevel: Assigns higher values to well-lit, low-traffic roads
    - EcoImpact: Assigns lower values to roads with a high environmental impact (for example, heavy emissions areas)

    Use the Calculate Field tool in ArcGIS Pro or the Field Calculator in QGIS to assign values to these fields.

3. **Configure your GIS analysis to favor or avoid certain segments during routing.**

    Set up your routing analysis to prioritize nicer routes instead of the shortest or fastest ones.

    In ArcGIS Pro, use the Find Best Route (Network Analyst) tool and set your cost attribute to ScenicScore instead of distance or travel time. Adjust the impedance settings so that low-scoring roads are less favorable.

    In QGIS, you need to do a little trickery. First, you need to create another field called ScenicSpeed. Then use the Field Calculator to assign ScenicSpeed to equal 1/ScenicScore (the inverse of the scenic score). This setting will ensure that higher-rated scenic roads have lower impedance, making them more favorable.

    Finally, use the QGIS QNEAT3 plug-in and select the Shortest Path (Point to Point) tool. Set the Optimization Criterion to Fastest and then choose your newly created field, ScenicSpeed, as the Speed Field. This method tricks QNEAT3 into prioritizing scenic routes by treating them as "faster" paths in the analysis.

REMEMBER

Nicest-path analysis is ideal for planning recreational trails, greenways, or scenic bike routes that prioritize user experience over speed and distance.

Defining service areas

How far can you go within a certain time or distance? This is the key question behind *service-area analysis*, which creates service areas. By creating and mapping these areas, you can use GIS to evaluate coverage, identify coverage gaps, and plan resource allocation. Here are some common uses for service-area analysis:

>> **Creating school bus routes:** GIS can create service areas based on the maximum number of passengers a bus can carry or the distance it can travel within a specified time frame.

- **Assessing fire station coverage:** Planners can determine which neighborhoods fall within a five-minute response time and pinpoint underserved areas that may need a new fire station.

- **Marketing campaign:** Businesses like newspapers or broadband providers can identify potential subscribers within their service area and target their marketing efforts.

- **Optimizing pizza delivery zones:** Pizza delivery services can make sure that their hot pizzas arrive on time by mapping delivery areas reachable within a set time frame.

Service area analysis not only determines coverage areas but also helps you to optimize resources, whether you're delivering pizzas or putting out fires.

Creating service areas requires a well-configured network database, typically the same one used for routing that includes impedances and turning rules. With that in place, you can use your GIS to create service areas. Figure 14-4 shows the results of a service area analysis for areas in a city that a driver can reach within five minutes.

Follow these steps to create a service area:

1. **Start from a central point.**

    This is the place for which you want to create a service area, like a school, fire station, or pizza joint.

2. **Expand outward along the network.**

    Your GIS software follows the network out from the central point, factoring in travel time, distance, impedances, and other restrictions.

3. **Generate a service polygon.**

    The result is a polygon showing reachable areas within your specified time and distance, such as a ten-minute drive or a five-mile radius.

To gain actionable insights for planning and decision-making, overlay demographic or infrastructure data on your service areas (see Chapter 15 for tips on working with map layers). This additional step will help you get a handle on the population served, businesses reached, or homes covered.

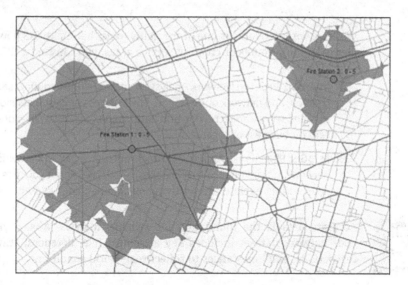

FIGURE 14-4: The results of a service area analysis.

Working with Networks in GIS

Both ArcGIS Pro and QGIS provide robust tools for network analysis, but getting good results depends on how well you prepare your data. You'll need a base network dataset to get started. Common network datasets include the following:

- **Local government data:** Many local and regional governments maintain detailed datasets that include road networks, traffic restrictions, and even real-time traffic updates. They are highly localized and often available for free via open data portals.

- **OpenStreetMap (OSM):** OSM is a globally crowd-sourced dataset that includes road networks, one-way restrictions, and turn attributes. Built by a dedicated user community, it's free and customizable.

- **State and regional transportation authorities:** Some states provide highly detailed transportation data, often with local road data integrated into it for seamless, statewide coverage.

- **StreetMap Premium from Esri:** StreetMap Premium is a robust and ready-to-use dataset. It includes turn restrictions, impedances, and travel speeds, but it comes at a price.

WARNING

When using datasets from outside sources, preprocessing is critical for accurate routing and service area results. Always validate the data's accuracy, review the quality standards, and check for missing attributes.

CHAPTER 14 **Working with Networks** 231

Working with networks is one of the many tools that make GIS powerful. Whether you're building routes or creating service areas, you can find a GIS app for that:

- » **In ArcGIS Pro:** The Network Analyst extension handles routing, service-area creation, and network optimization. The Business Analyst extension also includes service area creation and basic routing capabilities, tailored specifically for business applications.

- » **In QGIS:** Plug-ins like AequilibraE, Isochrones, and QNEAT3 provide flexible network analysis options that are similar to ArcGIS Pro. If you're feeling adventurous, explore advanced routing options using pgRouting with PostGIS.

Accurate data is the foundation for reliable analysis. Always make sure that your network dataset includes key attributes like distances, travel times, turn restrictions, and impedances to reflect real-world conditions.

> **IN THIS CHAPTER**
>
> » Selecting features with basic overlay methods
>
> » Combining attributes using spatial joins
>
> » Comparing overlaps with advanced overlay tools
>
> » Analyzing data with raster overlays

Chapter **15**

Exploring Map Overlay

Maps are tools for discovery. They're used to find places, plan cities, and manage natural resources. Every map starts with data. Whether that data relates to recreational trail networks, urban development, or soil erosion patterns, each map consists of layers that combine to tell a story. But how do you reveal that story?

One way to unlock the hidden insights within your maps is by using feature overlay tools. By comparing layers, you can find relationships between features, answer complex questions, and make data-driven decisions. Maybe you want to highlight businesses in a flood zone or see how land-use patterns overlap with zoning. Overlay techniques give you the tools to accomplish those tasks by turning raw data into meaningful insights.

In this chapter, you explore both basic and advanced overlay techniques and find out how they work in GIS software like ArcGIS Pro and QGIS.

TIP

This chapter gives you insights into using overlay functions in both ArcGIS Pro and QGIS. Get hands-on experience and see how they work by grabbing some sample data and trying these tools yourself! See Chapter 21 for some great data sources to get you started.

CHAPTER 15 **Exploring Map Overlay** 233

Exploring Basic Overlay Methods

Layering data is one of GIS's primary functions. An overlay tool lets you visualize different types of data together, like city points connected by road lines within state polygon boundaries. These layers are great for creating a state road map with major cities highlighted, but what if you want to compare the layers against each other? For example:

>> **Transportation:** How many miles of roads (lines) does a state (polygon) transportation department need to maintain?

>> **Water studies:** How many wells (points) are inside a particular county (polygon)?

GIS overlay tools help you analyze spatial relationships between different data layers. The basic methods described in this section, Select by Location and Spatial Join, are ideal when you need quick results or are prepping your data for further analysis.

TIP

When using overlay tools, most GIS software automatically adjusts the display of data layers that have different coordinate systems so that they align correctly on your map (see Chapter 2 for more about coordinate systems, projections, and spatial references). Even though this adjustment happens automatically in your GIS software, it's still good practice to check that all your layers have the correct coordinate system.

Selecting by location

When you need an answer to something like "What's near what?" or "What's inside what?" the Select by Location tool is the tool of choice. It's a quick and easy way to highlight (select) features that meet spatial criteria without creating a new layer or dataset.

For example, if you're working with a city planning office, you may be tasked with identifying all the businesses within a flood zone. You use GIS to select all the businesses (points) that fall within the flood zone (polygon). Now you can use those selected points to generate a quick report or simply visualize the results on a map to start thinking about your next steps.

Using this tool is straightforward, whether you're using ArcGIS Pro or QGIS:

>> **ArcGIS Pro:** Choose the Select by Location tool in the Selection group of the Map tab. Choose your input layer, specify the relationship (like "Within a

distance," "Contains," or "Intersects"), and run the tool to highlight the matching features.

» **QGIS:** Choose the `Select by Location` tool found on the project toolbar or from the `Vector` menu and go to `Research Tools > Select by Location`. Set your target and source layers, specify the relationship (like "Intersect," "Contain," or "Overlap"), and run the tool to select features.

TIP

Selecting features by location is great for highlighting features on the fly. But if you find yourself needing a new layer that combines attributes from overlapping or nearby features, use a spatial join instead.

Performing spatial joins

When you need to transfer attributes from one layer to another based on their location to each other, you want to use the spatial join tool. This tool creates a new layer with the combined attributes, which is a classic use of GIS.

For example, suppose you're studying burrowing owl nests. You have a point layer of the nests and want to add details about the parks (polygons) where the nests are located. A spatial join makes adding these details easy. By joining the park polygons to the owl nest points, you create a new point layer. In this new layer, each nest point includes the attributes for the park and where it's located, such as the park's name or size.

Here's how you perform a spatial join:

» **ArcGIS Pro:** Choose the `Spatial Join` tool found on the `Analysis` tab (or in the `Overlay` toolset of the `Analysis Tools` toolbox). Choose the target and join layers, define the relationship, and specify the output layer.

» **QGIS:** Choose the `Join Attributes by Location` tool, found in `Data Management Tools` on the `Vector` menu or `Vector General` in the `Processing Toolbox`. Choose the layers and the spatial relationship, and specify an output layer or create a temporary layer (the default).

TIP

When performing a spatial join, be sure you're using the right spatial relationship for the job. "Contains" and "Intersects," two of the most common spatial relationships, will give you different results, so choose carefully to make sure that your output layer has the connections you're looking for. If you're unsure, run both and compare the results to find the one that fits your needs.

CHAPTER 15 **Exploring Map Overlay** 235

Using Advanced Overlay Techniques

Overlay tools let you analyze how layers overlap — or don't overlap — and help you answer questions like "Which areas meet both of my criteria?" or "What's left when I remove overlapping areas?" These versatile tools work with points, lines, or polygons, allowing you to analyze spatial relationships with different types of data.

The examples in this section focus on common overlay tools available in most GIS software. Although you can apply these methods in both ArcGIS Pro and QGIS, their functionality sometimes differs a bit. Be sure to explore your software's documentation and play around with the different options. Who knows what stories your data may unveil!

Performing union overlays

A *union* overlay, as the name implies, combines all areas from two polygon layers into one output layer. The resulting output layer contains the attributes from both layers, making it a great tool for analyzing overlapping and non-overlapping areas together (see Figure 15-1).

For example, if you're working in urban planning, you may need to combine planning areas with flood zones to show all regions that are either planned for development, in a flood zone, or both. A union overlay creates a single dataset that includes all attributes from both input layers, ready for further analysis.

Here's how the process works across the two major desktop GIS platforms:

» **Feature types:** In ArcGIS Pro, union overlays support only polygons. QGIS supports points, lines, and polygons. Ideally, the geometry type of the input and overlay layers should be the same.

» **Output type:** The output is always polygon in ArcGIS Pro, but in QGIS the output type matches the geometry type of the input layers. For example, combining points and polygons will result in two separate runs for each geometry type.

» **Attributes:** The attributes from both layers are retained in the output.

» **Key difference:** QGIS's Union tool supports more geometry types, giving you greater flexibility, while ArcGIS Pro limits unions to polygon layers.

WARNING: Misaligned boundaries in your input layers can wreak havoc on your output layer, leaving you with *sliver polygons*, which are tiny polygon errors that don't reflect a real-world feature. To avoid such issues, use snapping or topology repair tools to clean your data before running the Union tool. (See the sidebar later in this chapter for tips on troubleshooting overlay issues).

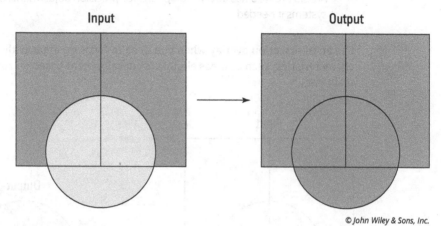

FIGURE 15-1: Union overlay.

© John Wiley & Sons, Inc.

Using intersection overlays

An *intersection* overlay retains only the areas where two polygon layers overlap. The resulting output layer contains the attributes from both layers but includes only the shared areas, making it a great tool for narrowing down your results to regions that meet multiple criteria (see Figure 15-2).

For example, suppose you're managing conservation efforts and want to find areas that are both in a conservation district and on public land. An intersection overlay keeps only the overlapping areas from both layers, along with their attributes, so that you can focus on just the areas that meet your criteria.

The Intersect tool works similarly in both ArcGIS Pro and QGIS (although it's called Intersection in QGIS):

- » **Feature types:** Both platforms support points, lines, and polygons as input or output geometry.
- » **Output type:** The output feature type is based on the lowest-dimensional type of input (with points being the lowest, followed by lines and then polygons).

CHAPTER 15 **Exploring Map Overlay** 237

- » **Attributes:** The attributes from all input layers are retained in the output layer.
- » **Key differences:** ArcGIS Pro requires you to specify an output feature class, but with QGIS you can create a temporary output for quick analyses. You can, however, use scripting tools in ArcGIS Pro if you want temporary results. ArcGIS Pro also has advanced options for precision adjustments or coordinate systems if needed.

TIP Use an intersection overlay when you need to focus on areas with multiple shared characteristics, such as areas eligible for development based on zoning and floodplain data.

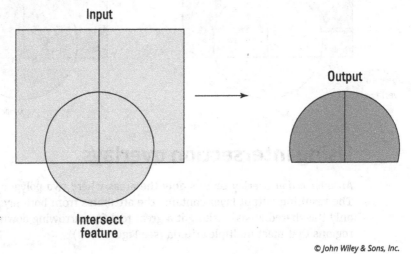

FIGURE 15-2: An intersection overlay of polygon layers.

© John Wiley & Sons, Inc.

Applying identity overlays

An *identity* overlay assigns attributes from one layer (called the "identity feature") to another, based on their spatial relationship. This tool is useful when you want to retain all features from your input layer while also incorporating attributes from an additional layer. The identity feature contributes its attributes to overlapping areas, but only where the two layers intersect (see Figure 15-3).

The output layer retains the full geometry and attributes of the input layer while adding the attributes from the identity feature wherever the two layers overlap. This combination allows you to analyze and visualize information from both layers in a single dataset, making it easier to draw meaningful conclusions.

For example, say you have a layer of potential development sites but need to know their zoning classifications, which are stored in a separate layer. By using the zoning layer as the identity layer, you can overlay it with the development site. The zoning attributes are then assigned to the development polygons, creating a new layer with overlapping portions that inherit zoning attributes.

Here's how the identity overlay works:

- » **Feature types:** In ArcGIS Pro, the input layer can be points, lines, or polygons, but the identity layer must be a polygon or the same type as the input (such as points to points or lines to lines).
- » **Output type:** Matches the geometry type of the input layer.
- » **Attributes:** The attributes of both layers are retained in the output layer.
- » **Key difference:** QGIS doesn't have a dedicated identity tool. You can, however, combine the Intersection tool in QGIS with field calculations or spatial joins to get similar results.

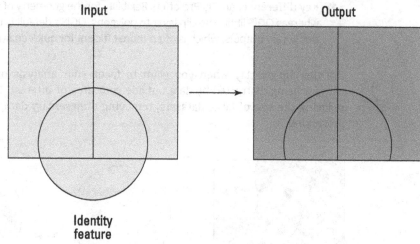

FIGURE 15-3:
An identity overlay.

© John Wiley & Sons, Inc.

Applying clip overlays

A *clip* overlay slices one layer to match the boundaries of another. It's like using one layer as a cookie cutter over another because it keeps only the parts of the input layer that fall within the clip layer's boundaries (see Figure 15-4).

CHAPTER 15 **Exploring Map Overlay** 239

In Figure 15-4, the input layer (left) contains various features, including contour lines and points. The clip feature (the circle in the middle) acts as the cookie cutter, defining the area to keep. The output layer (right) keeps only the portion of the input layer that falls within the clip feature, removing everything outside its boundary.

For instance, say you're working on a vegetation study and want to focus solely on a specific study area. A clip overlay trims your vegetation layer to the boundaries of the study area polygon.

Here's how a clip overlay works in ArcGIS Pro and QGIS:

» **Feature types:** In ArcGIS Pro, both the input and clip layers can be points, lines, or polygons. However, if the input feature is polygon, then the clip feature must be polygon. And, if the input features are lines, the clip features can be lines or polygons, not points. In QGIS, the clip layer must be a polygon.

» **Output type:** Matches the input layer's geometry type.

» **Attributes:** The output layer retains attributes only from the input layer.

» **Key difference:** ArcGIS Pro offers flexibility for the geometry of your clip layer, whereas QGIS limits the clip layer to polygons. QGIS's default is to create temporary outputs, which is often more efficient for quick analyses.

TIP

Use the clip overlay when you want to focus your analysis on a specific area without being distracted by data outside your area of interest. It's also great for reducing the size of large datasets, removing unnecessary data, and speeding up processing.

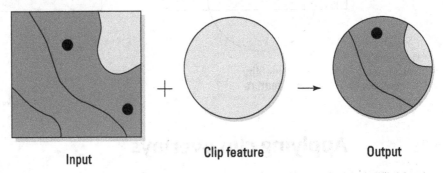

FIGURE 15-4: Clip overlay.   Input   Clip feature   Output

© John Wiley & Sons, Inc.

240   PART 4 **Analyzing Geographic Patterns**

Exploring symmetrical difference overlays

A *symmetrical difference* overlay highlights areas where two polygon layers don't overlap. This tool is useful for identifying mismatches, gaps, or inconsistencies between datasets (see Figure 15-5). It's especially helpful for finding regions that overlap or locating areas excluded from both datasets, such as shared boundaries that don't align perfectly.

For example, if you're analyzing wildlife habitats and protected areas, you may need to target restoration efforts in spots that are not protected and not identified as a current habitat area. A symmetrical difference overlay can help pinpoint the regions falling outside both zones, allowing you to quickly find your target areas. Along the same lines, for infrastructure projects, you can compare approved development zones against completed projects, highlighting gaps or discrepancies for follow-up.

Here's how the symmetrical difference overlay works in ArcGIS Pro and QGIS:

» **Feature types:** Both platforms support points, lines, and polygons as input, and the output does not have to match the input geometry type.

» **Output type:** The output type is determined by the lowest-dimensional input (with points being the lowest). For instance, combining a line and a polygon layer will result in line features for the output.

» **Attributes:** The output layer retains attributes from both input layers.

» **Key difference:** In QGIS, the default output is a temporary layer, allowing you to quickly and easily explore data without creating new layer files. ArcGIS Pro requires saving the output as a permanent layer, though you can use custom scripting to get around that limitation.

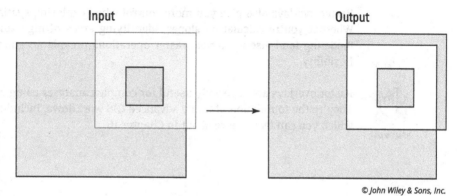

FIGURE 15-5: Symmetrical difference overlay.

© John Wiley & Sons, Inc.

TIP

You use the symmetrical difference overlay tool to clean and validate your data, ensuring alignment between layers and uncovering areas that might otherwise go unnoticed. This method is a good approach when you need to focus on the differences rather than the similarities between two datasets.

Understanding Raster Overlay

Raster overlay tools allow you to do many of the same things that you can do with vector — point, line, and polygon — overlays, but with added flexibility. If your analysis involves comparing continuous data like elevation, temperature, or land cover, raster overlay can save you time and frustration — a real life saver!

Here's what makes a raster overlay different from a vector overlay:

>> **Grid-based precision:** Each cell in a raster has a unique value, and every cell aligns perfectly with corresponding cells in other layers. This alignment (called *coregistration*) eliminates issues like mismatched edges or sliver polygons that can complicate vector overlays.

>> **Mathematical and statistical power:** Raster overlay allows you to compare layers using mathematical expressions, logical conditions, or statistical measures, making it ideal for detailed spatial modeling.

Suppose you're creating a suitability map for a new wildlife corridor. You can overlay raster layers for elevation, vegetation density, and proximity to roads and then use mathematical expressions to assign weights to each factor. Assign higher weights to vegetation density and lower weights to proximity to urban areas, reflecting their relative importance.

Raster overlays also give you more control when exploring spatial relationships. Whether you're calculating slopes, identifying areas of high solar potential, or modeling land-use scenarios, raster operations provide an amazing amount of flexibility.

REMEMBER

Raster overlays are especially useful for complex analyses using continuous data. They're the foundation of many advanced GIS workflows, including map algebra, which you can learn more about in Chapter 16.

TROUBLESHOOTING OVERLAY ISSUES

Even the most well-prepared data can run into hiccups sometimes. This sidebar addresses some common issues you might run into when overlaying data.

Sliver polygons (vector data only)

These are small, unwanted polygons that don't reflect the real world. Slivers happen when your input layers don't align perfectly, causing gaps or tiny overlaps in the output layer. Try these fixes:

- **Topology tools:** Clean up polygon boundaries by using topology validation tools in ArcGIS Pro (like the `Fix Topology Error` tool) or QGIS (try the `Snap Geometries to Layer` tool).
- **Snapping:** Enable snapping during editing to ensure vertices from different layers align properly.
- **Buffering:** For persistent slivers, apply a small buffer (positive or negative) to clean up overlaps or gaps before running an overlay.

Empty results (vector and raster data)

Unexpected empty output can happen if your input layers don't spatially overlap, or the spatial relationship you choose (like `Contains` or `Intersects`) doesn't apply to your data. Try these fixes:

- **Check spatial overlap:** Zoom into your layers and make sure they overlap where you expect them to.
- **Adjust spatial relationships:** Test different relationships to find the correct one for you analysis. For example, try `Within` instead of `Intersects`.
- **Expand input layers:** If your input boundaries are too restrictive, apply a buffer to one or both layers to ensure that features from different layers have overlapping areas.

(continued)

(continued)

Coordinate mismatches (vector and raster data)

When your layers look misaligned or the results seem inaccurate, you may be mixing coordinate systems. Try these fixes:

- **Project layers:** Use tools like `Project` in ArcGIS Pro or `Reproject Layer` in QGIS to ensure that all the inputs share the same coordinate system.

- **Verify on-the-fly projection:** Check to see whether your software is projecting the layers correctly. If your results still seem off, manually project the layers and then run the overlay tool.

- **Choose a consistent datum:** For high-precision overlays, ensure that the layers share the same datum, such as NAD83 or WGS84.

IN THIS CHAPTER

» Using cartographic modeling

» Getting the hang of map algebra

» Building and refining cartographic models

Chapter 16
Mastering Map Algebra and Cartographic Models

If you're ready to see how GIS can spread its wings and do something really spectacular, this chapter is for you! Although retrieving data and running basic analyses gets the job done, sometimes you'll want to bring multiple datasets and techniques together in complex ways by using techniques like those described in this chapter, cartographic modeling and map algebra. These are some of the more powerful GIS tools and concepts that you can use to transform raw data into meaningful insights and make your GIS really soar.

In this chapter, you find out how to design and implement cartographic workflows, use map algebra to combine and analyze raster data, and test your results for accuracy to ensure that user needs are met. Whether you're identifying suitable sites for solar panels or modeling flood risks, these techniques open the door to more advanced GIS applications.

THE ORIGINS OF CARTOGRAPHIC MODELING

The original ideas and concepts in this chapter are the brainchildren of Dr. C. Dana Tomlin and his PhD advisor, Dr. Joseph K. Berry, then at the Yale School of Forestry. Their groundbreaking research in the 1980s helped to shape the field of cartographic modeling and remains foundational to GIS modeling today.

Concerned with more than just creating maps, cartographic modeling also involves developing workflows that combine data, tools, and logic to solve real-world problems. Dr. Tomlin's work on map algebra, a language for combining and analyzing raster layers, became a cornerstone of GIS and inspired the tools that many professionals use today in software like ArcGIS Pro and QGIS.

Creating Cartographic Models

The simplest definition of *cartographic modeling* is that it combines spatial data, GIS functions, and operations to generate new information. Cartographic models are workflows that use GIS tools to analyze data, answer questions, and support decision-making. With these models, you combine common GIS operations like reclassification, buffering, overlay, and interpolation to create results tailored to your specific project needs. In fact, if you've done any sort of geospatial analysis, you may have already used a cartographic model without realizing it.

I like to think of cartographic models as the nerdy part of the decision-making process, transforming raw data into useful information for various types of analyses, including:

- **» Suitability analysis:** Combine data on lakes, protected areas, and road access to figure out the best locations for new parks. Are the sites near lakes? Away from protected areas? Accessible by road?

- **» Risk assessment:** Determine areas vulnerable to flooding using data on rainfall, land use, and elevation.

- **» Urban planning:** Rank parcels for their development potential based on proximity to utilities, zoning rules, and transportation access.

Cartographic models aren't just for GIS nerds. By helping make complex spatial decisions easier to apply, you find them applied in the following ways in areas like these:

- **Public health:** Tracking disease outbreaks and planning responses
- **Transportation studies:** Designing optimal routes and corridors
- **Site selection:** Choosing locations for a new business or school
- **Conservation:** Protecting endangered species and habitats

When building cartographic models, think about the questions you want to answer and let that guide your choice of data, tools, and methods. Tools like ArcGIS Pro ModelBuilder or QGIS Model Designer (covered later in this chapter) can help you design and automate your workflows.

Understanding Map Algebra

Some people like algebra; some don't. If you're among the latter, just the title of this section may send chills down your spine, but fear not: Map algebra is simply a way of combining raster layers to analyze and create new information. (Refer to Chapter 8 for the basics of raster data). Each cell in a raster layer has a value, representing elevation, temperature, or land cover type. Map algebra allows you to perform calculations using the values from multiple raster layers to answer questions and solve problems.

Map algebra applies the math and logic to one or more raster layers. Here's what you can do with map algebra:

- **Sum two layers** to calculate the total rainfall over a given period or combine different environmental factors, such as soil moisture and vegetation density, to assess drought conditions.
- **Subtract one layer from another** to find the difference in elevation between a Digital Surface Model (an elevation raster that includes vegetation, buildings, and other objects) and a Digital Elevation Model (a raster representing the elevation of the bare ground, excluding vegetation, buildings, and other objects).
- **Classify cells into categories** to highlight areas that meet specific criteria, like slopes suitable for construction.

The beauty of map algebra is that it's built on the simplicity of raster grids. Each cell lines up perfectly across layers, so combining data is simple and straightforward. Map algebra is also versatile and easy to scale up from simple calculations

CHAPTER 16 **Mastering Map Algebra and Cartographic Models** 247

to complex models involving multiple layers. For example, if you're using GIS to plan a hiking trail, you might follow these steps:

1. Start with a slope layer to identify flat areas.
2. Add a land-use layer to avoid building a trail across protected areas.
3. Add a proximity layer to ensure that the trailhead is close to parking areas.

Using map algebra, you combine these layers to create a suitability map showing the best areas for the trail. Modern GIS software, like ArcGIS Pro and QGIS, makes map algebra user friendly with tools like the Raster Calculator shown in Figure 16-1. In the next section, I walk you through some of the most common map algebra functions.

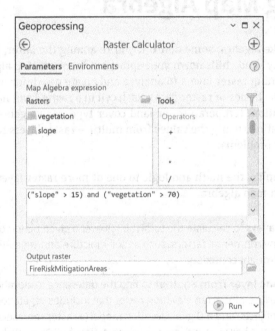

FIGURE 16-1: The Raster Calculator in ArcGIS Pro.

Performing Map Algebra Functions

Map algebra powers a lot of the geospatial analyses in GIS, from simple calculations to more involved workflows. It helps you combine raster layers, apply mathematical operations, and create new spatial insights. Whether you're identifying areas suitable for a hiking trail or mapping flood risks, map algebra can help you turn raw data into maps, reports, and data dashboards that provide valuable insights.

Exercising control in cartographic modeling

Modern GIS platforms like ArcGIS Pro and QGIS make map algebra much more accessible than in the early days of GIS, when it required tedious command-line operations or custom programming. Ready-to-use tools like the Raster Calculator (refer to Figure 16-1 in the previous section) walk you through the process, allowing you to

- » **Choose inputs:** You can add raster data layers, tables, or constants.
- » **Define operations:** You use math operators (+, −, *, /) and logical expressions (<,>, AND, OR) to combine data.
- » **Visualize results:** You can see the output of your operations as you perform them.

As an example of using map algebra, here are steps you might take to map fire risk for an area:

1. **Add rasters for slope and vegetation density.**
2. **Use this formula: (Slope > 15) AND (Vegetation > 70).**

   The output highlights steep slopes and dense vegetation, which are important factors in assessing fire risk and planning mitigation efforts.

TIP

The use of logical operators, like the ones that appear in the preceding Step 2, simplifies complex queries. When you need to focus on specific conditions such as finding areas with a steep slope AND dense vegetation, logical expressions work great.

Using local functions

Local functions are the simplest type of map algebra because they work on individual cells without considering neighboring ones. Each cell in the output is calculated independently based on one or more input layers.

Figure 16-2 shows an example of how a local function works, which is by adding two rasters together, cell by cell. Notice how each cell in the output is the sum of corresponding cells from the input rasters; for example:

- » The top leftmost cell in the first raster is 8, and the top leftmost cell in the second raster is 1. Adding the two cells together, the output cell in the top leftmost position becomes 9.

>> This operation is repeated independently for each of the corresponding cells in the raster grids.

FIGURE 16-2: An example of a local function, with two rasters being added together on a cell-by-cell basis.

© John Wiley & Sons, Inc.

Table 16-1 lists some common types of local functions and what you do with them. The functions listed in the table are primarily for ArGIS Pro's Raster Calculator and Map Algebra. Many of the functions also work in QGIS, but QGIS often relies on simpler mathematical expressions instead of named functions.

TABLE 16-1 Common Types of Local Functions

Type of Function	Functions	Use	Example
Selection	Con(), SetNull()	Use logical conditions to extract specific values.	Find all cells where land use equals "forest" and elevation > 500 meters.
Trigonometric	Sin(), Cos(), Tan(), Atan()	Perform calculations like sine and cosine.	Useful for advanced mathematical or terrain analysis like modeling solar radiation.
Exponential and logarithmic	Exp(), Log(), Sqrt(), Power()	Handle transformations like squares, cubes, and log base 10.	Often used in scaling or normalizing data.
Reclassification	Reclassify(), Remap()	Change cell values based on rules.	Group elevation ranges into categories like "low," "medium," and "high."
Statistical	Mean(), Median(), Mode(), Sum()	Apply operations like mean, median, and mode.	Great for summarizing numerical data in each cell.
Other math	Plus(), Minus(), Times(), Divide()	Perform arithmetic, like addition or subtraction, or do conditional comparisons.	if slope > 15, return 1; else, return 0.

In QGIS, you can perform many of the functions listed in Table 16-1 using basic mathematics, though some named functions are also available. For example, arithmetic operations use +, -, *, and /, whereas logical conditions can be written using if() statements (for example, if(slope > 15, 1, 0)).

Local functions are useful for quick, cell-by-cell operations, like classifying terrain into high and low elevations or calculating the difference between two temperature layers.

Using neighborhood functions

In map algebra, a *neighborhood* is a small, defined area around a cell that influences its value in raster analysis. Neighborhoods are central to *neighborhood functions*, also called focal functions, which analyze each cell in relation to its neighbors rather than individually. These functions help identify patterns, smooth data, or assess how surrounding values influence a particular location.

A neighborhood function works like this: instead of analyzing just one cell in isolation, a neighborhood function applies a moving window, often in the form of a 3-x-3 grid, centered on each cell and performs a calculation based on the values in that window.

To better explain this concept, take a look at Figure 16-3, which illustrates the maximum function. The *maximum function* is a neighborhood function that identifies the highest value in a 3-x-3 neighborhood and assigns that maximum value to the center cell in the output raster.

Here's how the maximum value works:

1. **Identify the maximum value in the 3-x-3 neighborhood.**

   In the left grid of Figure 16-3, it's the bold 2 in the center cell of the current 3-x-3 neighborhood. The highest value is this neighborhood is 8.

2. **Assign the maximum value to the center cell in the output raster.**

   In the corresponding output raster (the grid on the right), the center cell's value (previously 2 in the input raster) is updated to 8, the highest value from the neighborhood (bold 8 in the right grid of Figure 16-3).

3. **Repeat for every cell.**

   The moving 3-x-3 neighborhood passes over the entire input raster, evaluating each 3-x-3 neighborhood for the maximum number and placing that maximum number in the corresponding output raster cell.

CHAPTER 16 **Mastering Map Algebra and Cartographic Models** 251

FIGURE 16-3:
The maximum function replaces the center cell's value with the highest value from its 3-x-3 neighborhood.

© John Wiley & Sons, Inc.

Defining your neighborhood: shapes and sizes

The shape and size of a neighborhood function affect how spatial relationships are analyzed. Some common neighborhood shapes include the following:

» **3-x-3 square:** The most basic neighborhood that includes the center cell and its eight surrounding neighbors, forming a 3-x-3 grid (refer to Figure 16-3). This type of neighborhood is useful for general-purpose operations like smoothing or identifying outliers.

» **Circle:** This is a circular neighborhood that includes all cells within a specified distance from the center, making it ideal for distance-based analyses, like modeling the spread of sound, fire embers, or pollution (see Figure 16-4).

» **Wedge:** The wedge neighborhood focuses on a specific direction, useful for simulating water flow or wind-driven fire spread (see Figure 16-4).

» **Annulus (Donut):** This neighborhood analyzes a ring of influence, excluding the center cell, ideal for studying edge effects or concentric patterns like vegetation zones around a water source or deforestation radiating outward from a city (see Figure 16-4).

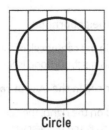
Circle

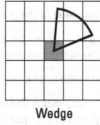
Wedge

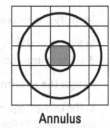
Annulus

FIGURE 16-4: Common neighborhood function types.

© John Wiley & Sons, Inc.

When using neighborhood functions, if you're calculating the spread of wildfire embers, a circle neighborhood simulates how far embers might travel, whereas a wedge can represent fire spread in a specific direction because of wind. Meanwhile, an annulus can help evaluate edge effects, such as deforestation radiating outward from the urban centers.

Choosing the right neighborhood depends on the question you're asking. For example:

>> **How do nearby cells affect the center cell?** Use a 3-x-3 square.

>> **What's the pattern of influence around a point?** Try a circle or annulus.

>> **How do directional forces like wind affect an area?** Go for a wedge.

TIP

Choose the shape and size of your neighborhood based on what you're trying to analyze. For example, if you're modeling noise levels from a busy road, a circle might make sense. But if you're analyzing square blocks in a city, a square neighborhood might fit better.

Making sense of neighborhoods in action

Earlier in this section, "Using neighborhood functions," I explain how neighborhood-based calculations work: a moving window, often a 3-x-3 square, slides across a raster, analyzing each cell in relation to its neighboring cells. One common example of this neighborhood function in action is smoothing temperature data — your GIS software calculates the average temperature within each neighborhood (the moving window) and assigns that value (the average temperature) to the corresponding cell in the output raster. This process reduces data noise and highlights broader trends in the data.

These neighborhood shapes aren't just theoretical — they're designed to reflect real-world scenarios. Here are some examples of how different shapes fit specific uses:

>> **Squares and Rectangles:** Best for human-made structures like city blocks, agricultural fields, or parking lots. While squares (like 3-x-3 grid neighborhoods) area most common, some tools allow rectangular neighborhoods (like 3-x-5 grid neighborhoods) for specific analysis.

>> **Circles:** Compact and commonly used for analyzing natural phenomena like noise, flood zones, or wind impacts.

>> **Wedges:** Ideal for directional data like wind flow or river deltas.

>> **Annulus (Donuts):** Useful for isolating edge effects, like detecting urban boundaries.

In GIS software like ArcGIS Pro and QGIS, you can define neighborhoods through GIS tools or custom scripts, allowing you to tailor the shape and size to your needs. Tools like the Focal Statistics raster function (see Figure 16-5) in ArcGIS Pro's Image Analyst and Spatial Analyst extensions let you preview and adjust these settings interactively. For example, you can specify a rectangle, circle, or annulus as the neighborhood type and experiment with different statistics, such as mean, maximum, or standard deviation.

TIP

Experiment with different neighborhood shapes and functions to see what fits your data best. Try combining smaller neighborhoods for localized smoothing with larger ones for broader trends. You might uncover patterns you didn't expect!

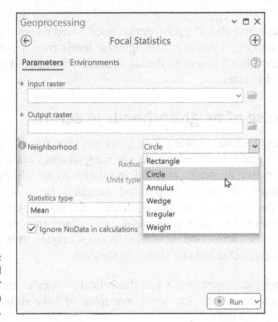

FIGURE 16-5: The Focal Statistics raster function tool in ArcGIS Pro.

Exploring zonal functions

Zonal functions take a different approach than neighborhood (focal) functions. Instead of analyzing a single cell and its neighbors, zonal functions group cells into zones based on shared attributes or geometry and then analyze each zone as a whole. Imagine zoning a city into districts, like residential, commercial, or industrial areas, and then analyzing each district as a whole. That's the essence of zonal functions.

Zones are groups of grid cells that share the same value or characteristic, forming meaningful regions for analysis. A zone can be a single contiguous region or multiple disconnected areas that share the same classification. Here are some examples of zones:

» All the cells representing a specific land-use type, like forest (even if forested areas are scattered throughout the map)

» Elevation ranges grouped into lowlands, midlands, and highlands

» Administrative regions like counties or school districts

After you define zones, zonal functions let you compare, calculate, and summarize values for each zone, regardless of whether the areas are contiguous or scattered.

Putting zonal functions to work

Zonal functions follow a simple, three-step process:

1. **Define your zones.**

   Your zones can come from a vector feature layer, a single raster layer, or a combination of layers. For instance, you might use county boundaries or define zones based on land cover type (wetlands, forest, desert, and so on).

2. **Specify the data to analyze.**

   Use a raster layer containing the values you want to analyze, like rainfall or temperature. See Chapter 8 for how to work with raster data.

3. **Perform the analysis.**

   Use zonal raster tools in GIS software like ArcGIS Pro or QGIS to calculate statistics or other measures for each zone.

For example, suppose you're studying biodiversity across a national park. Your zones can be forest types (pine, oak, and mixed forest), and your data layer might represent species counts. A zonal function can calculate the average species richness for each forest type. The resulting table, like the one shown in Figure 16-6, helps ecologists target conservation efforts.

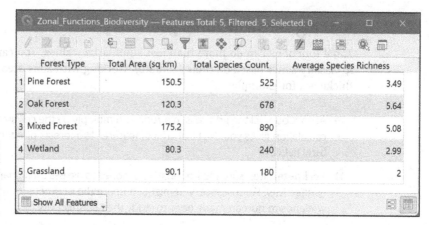

FIGURE 16-6: Zonal functions use zones to summarize data across an entire area.

CHAPTER 16 Mastering Map Algebra and Cartographic Models 255

Using common zonal operations

Zonal functions can handle a variety of tasks, from crunching numbers to analyzing shapes. Here are some of the most common types of zonal functions and examples of how you can use them:

- **Statistics:** Calculate values like average (mean), total (sum), or range (difference between maximum and minimum) for each zone. You might use this to find the average elevation for each watershed.

- **Majority and variety:** Determine the most common value (majority) or the number of unique values (variety) in a zone. For example, ecologists might identify the zone with the greatest variety of habitat types to pinpoint areas of high biodiversity.

- **Geometry-based metrics:** Calculate geometric properties like area, perimeter, or thickness of zones. You can use this method to measure the area of each agricultural field or the thickness of riparian buffer zones along rivers.

Zonal functions are incredibly useful and versatile across various fields. Here are some ways to use them:

- **Environmental analysis:** Identify zones with high rainfall to assess flood risk.

- **Urban planning:** Measure population density in different districts to help plan public services.

- **Agriculture:** Calculate the average soil quality in each field to guide crop rotation.

- **Conservation:** Analyze habitat variety in protected areas to prioritize conservation efforts.

Making sense of zonal geometry

Zonal functions can do more than just summarize values. You can also use them to classify zones based on their geometry, working with zone area, perimeter, and thickness, for example:

- **Area:** Count all the grid cells in a zone and multiply by the cell size. This calculation gives you the total area of features like forests or agricultural fields.

- **Perimeter:** Measure the boundaries of zones to assess their shape and connectivity. For instance, you might compare the perimeter of natural wetlands to human-made ones to study their connectivity.

» **Thickness:** Analyze the width of zones to identify long, narrow features, like mountain ridges or river corridors.

Resampling raster data

You don't have to work with raster data for long before you come across datasets with different resolutions, especially if you're using data from a variety of sources — as is often the case. Varying resolutions can cause all kinds of compatibility issues, so to save yourself frustration, you need to resample the data. *Resampling* adjusts cell sizes, either by aggregating or interpolating the data to create a new dataset with uniform resolution.

Here are some reasons you may need to resample raster data:

» **Analysis:** For meaningful analysis, datasets with different resolutions need to align. Misaligned data can lead to errors or misleading results.

» **Efficiency:** Generalizing data through resampling reduces file size and processing time, making workflows more manageable.

» **Modernization:** Resampling allows you to update older datasets to match the spatial resolution of newer or higher-quality data.

Suppose you're working on a project and need to compare land cover from a 20-meter resolution raster from 1951 with a 10-meter resolution raster from 2025. To ensure compatibility, you need to resample the 1951 data to a 10-meter resolution using the majority method in ArcGIS Pro or the mode method in QGIS (look for it in the Warp (Reproject) tool). Using one of these methods ensures your analysis accurately captures patterns and avoids errors caused by misaligned datasets.

Resampling uses specific techniques to calculate new call values. Common methods, which are available in modern GIS software, include

» **Nearest neighbor:** Copies the value from the nearest original cell. This method is fast but less accurate for continuous data like temperature or elevation.

» **Bilinear interpolation:** Averages values from the nearest four cells, making it better suited for continuous data.

» **Cubic convolution:** Uses a weighted average from a larger neighborhood, producing smoother results but requiring more processing time.

» **Aggregation-based methods:** Groups cells into larger blocks and applies functions like Mean, Median, Mode (Majority), Sum, or Maximum. For example, you might use Mean aggregation to simplify rainfall data for regional trends.

To change resolution or align datasets in ArcGIS Pro, use the Resample tool, found in the Data Management toolbox after choosing Raster ⇨ Raster Processing. In QGIS, use the Warp (Reproject) tool available in the Processing Toolbox (choose GDAL ⇨ Raster Projections).

Both ArcGIS Pro and QGIS resampling tools allow you to select a resampling method within their resampling tool options. The resampling method you choose can significantly impact your analysis, so it's important to match it to your data type and analysis needs. For example, for categorical data such as land use, nearest neighbor works well, whereas continuous data benefits from interpolation techniques. Test different methods on a small subset of data to see how they impact your analysis.

Using global functions

Global functions take a bird's eye view of your entire study area. Unlike local or neighborhood functions, which focus on individual cells or smaller areas and are described in the previous sections, global functions operate across the whole dataset simultaneously. Global functions are very powerful and complex operations, but they're perfect for analyzing large-scale patterns and relationships.

Here are some common examples of global functions and how you can use them:

» **Distance measures:** Calculate the distance from a specific location to every other cell. Here are some distance calculation methods:

- **Euclidean distance:** Straight-line distance, like "as the crow flies"
- **Manhattan distance:** Distance following a grid, like walking through city streets

See Chapter 11 for more on measuring distance.

» **Surface functions:** Model terrain features such as basins, pour points, and drainage networks to understand how water flows across landscapes. See Chapter 12 for details on working with surfaces in GIS.

» **Interpolation functions:** Predict values at unsampled locations by using known data points. Techniques include linear, nonlinear, trend surface, and kriging. (Chapter 12 explains interpolation techniques.)

» **Hydrology functions:** Model water accumulation, flow direction, and drainage patterns to analyze watersheds or flood risks.

These operations are ideal for scenarios where individual cell analysis isn't enough, and you need to understand how the entire dataset interacts as a whole.

Exploring global function groups

Some global functions dive even deeper into specific (and very geeky) applications, like groundwater modeling and multivariate statistics.

The groundwater global functions are a specialized category of global functions designed to analyze the flow of liquids underground. These tools help model how water moves through permeable materials like sand or gravel and track the spread of dissolved substances (like pollutants). They're commonly used for

» Mapping underground aquifers

» Simulating water movement based on slope and pressure (known as *head gradient*)

» Tracking pollutants from point sources (like a leaking pipe) or non-point sources (like agricultural runoff)

If you're working in hydrology, these functions are incredibly powerful. Specialized hydrology software might offer even more robust tools. In ArcGIS Pro, the Spatial Analyst extension has toolsets for both groundwater and hydrology analyses. In QGIS, you'll need to search the processing toolbox for available options.

TIP

Although QGIS doesn't have dedicated toolboxes like ArcGIS Pro's Spatial Analyst extension, you can get a wide array of hydrology and groundwater modeling tools through plug-ins and integrations. Check out the GRASS GIS and SAGA GIS plug-in toolboxes for watershed delineation, flow accumulation, and stream network extraction. The WhiteboxTools plug-in is another excellent option, offering lightweight hydrological analysis. GRASS GIS is a core plug-in but you'll most likely need to download, install, and configure SAGA GIS and WhiteboxTools to take advantage of their many tools.

Mixing statistics and GIS

Multivariate functions apply advanced statistical techniques across multiple raster layers to analyze complex relationships. While these functions aren't exclusive to

GIS, they're built into most professional GIS tools like ArcGIS Pro and QGIS. Here are some examples of how you can use multivariate functions:

» Combine layers like soil quality, rainfall, and crop yield to find patterns in agricultural productivity using functions such as Principal Component Analysis (PCA) or Canonical Correlation Analysis (CCA). These functions are available in ArcGIS Pro's Spatial Analyst extension and QGIS via the SAGA NextGen plug-in.

» Use statistical techniques like Regression Analysis (Ordinary Least Squares, Geographically Weighted Regression) or K-Means Clustering to model relationships across multiple variables or clustering to model relationships across multiple variables. You can find these functions in ArcGIS Pro's Spatial Statistics toolbox and QGIS through either the SAGA NextGEN plug-in or GRASS GIS plug-in.

Although these functions can get quite technical, they're great for exploring patterns and correlations that aren't obvious at first glance. Here are some areas where you might find these functions used in real-world applications:

» **Urban planning:** Use Euclidean (straight-line) distance to calculate service areas for schools, hospitals, or fire stations.

» **Environmental management:** Model drainage networks to plan stormwater systems or identify areas prone to flooding.

» **Natural resource management:** Predict groundwater flow to assess potential contamination risks or identify water sources.

» **Climate modeling:** Perform large-scale interpolations to estimate temperature or precipitation patterns in unsampled regions.

Building a Model

Whether you use map algebra or any other GIS tools, your analysis will likely involve multiple datasets along with a variety of tools and techniques. Combining these into logical steps is what GIS folks call *model building*. This section walks you through the process of planning your model, preparing data, and implementing a GIS model from start to finish.

Planning your analysis

Whether you're going on a vacation or baking cookies, you typically work with a plan. Every model of an activity begins with a plan outlining the steps needed to complete the task at hand. Although plenty of software tools are available for you to use to outline your steps, I find the best place to start is to simply go analog — in other words, grab a pen and paper! Start by diagramming the steps needed to complete your analysis, as I've done in Figure 16-7. I find it easiest to start by writing the desired outcome across the top of the page, so I always have that in mind. Then break that down into smaller pieces like the criteria and data needed.

To show you how this might look, take a look at Figure 16-7. Here I outline the basic requirements for developing a map of ideal sites for a hypothetical solar farm called Solsites. The ideal sites need to meet these criteria:

» Each site must be at least two miles from town.

» Each site should be on slopes between 30 and 55 degrees.

» The slopes must be south-facing for optimal sun exposure.

REMEMBER

Getting this information down on paper helps to keep the steps organized, ensure that all criteria are included, and even identify data needed to complete the process. Essentially, it's a visual diagram of your model. You can, of course, do something similar using a spreadsheet, bullets in a text document, or notes scribbled on a whiteboard. Use whatever works best for you; the point is to roughly outline your process before jumping into the modeling tools.

TIP

Keep your outline handy so that you can refer to it as you create your model. If additional criteria come into play or things change, you can always modify your draft outline. If you're pretty sure that changes will be needed, try using a whiteboard so that you can easily erase and modify as needed.

Notice how, in Figure 16-7, I divide the chart by the three requirements for an ideal site location: good slope, good aspect, and good distance from town. Then I list the specific requirements for what makes a site "good" along with the data I will need in order to find my ideal site.

REMEMBER

By laying out your process with a basic outline, you not only jump-start your "plan of attack" for the project but also identify your data needs to complete it. At this point, don't worry about whether the data you need are available; you'll work on that in the next step. Here, just concentrate on the overall process.

CHAPTER 16 **Mastering Map Algebra and Cartographic Models** 261

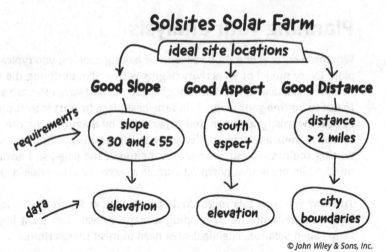

FIGURE 16-7:
A draft outline of the requirements to create the Solsites map.

Preparing your data inputs

After you lay out your plan for a project, such as locating the ideal site to build a solar farm, it's time to gather and prep the data. Gathering and prepping your data will most likely include the following:

- » **Identifying and downloading datasets**, such as elevation, land use, or city boundaries
- » **Cleaning and reclassifying data** to match your criteria
- » **Aligning datasets** by resampling rasters to the same resolution or projecting layers to the same coordinate system

TIP

Keep track of the tools you use in your data prep. If you know you'll be using these workflows again in the future, you can incorporate the data preparation methods into your model or even build stand-alone models for cleaning and prepping the data for use in other projects.

Implementing your model

Modern GIS software makes model implementation accessible by providing tools like ArcGIS Pro's ModelBuilder and QGIS's Model Designer, which allow you to drag and drop processes into a flowchart-like interface. With these tools, you can build a sequence of steps, define parameters, and even add descriptions for each process to document your model's purpose and logic. The saved model can be shared, reused, and adapted by others.

For example, take a look at the model for the hypothetical Solsites project in Figure 16-8. It's built in ArcGIS Pro's ModelBuilder and includes the input digital elevation model (DEM), slope and aspect calculations, along with buffered city boundaries — all in one automated workflow. The output of this model is a raster layer that highlights the areas that meet the Solsites criteria for slope, aspect, and distance from cities.

TIP

Automated modeling tools save time and reduce errors. Plus, with a drag-and-drop interface, they let you easily tweak and repeat your analysis as needed, which is great for refining results or experimenting with different criteria. Try it on your next project, and you can thank me later for how much time it saved you!

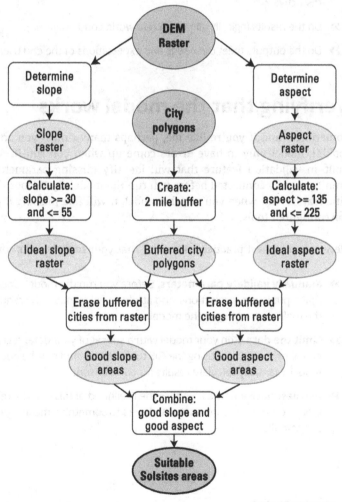

FIGURE 16-8:
A model in ArcGIS Pro for the Solsites project.

© John Wiley & Sons, Inc.

CHAPTER 16 **Mastering Map Algebra and Cartographic Models** 263

Testing a Model

No model is perfect. Even the best-designed models need testing to ensure that they work as intended. Testing is a crucial part of the modeling process and involves checking to see whether your model produces realistic and reliable results.

Testing your model comes down to posing four basic questions, which I cover in more detail in the following sections:

- » Does the model run correctly, producing the expected results without errors?
- » Are the model's outputs appropriately specific (not too broad or too restrictive)?
- » Do the results logically align with real-world conditions?
- » Do the outputs meet the needs and expectations of the end user?

Verifying that the model works

Sometimes (and if you're like me, perhaps many times) you create a beautiful-looking model only to have errors come up when you run it. ArcGIS Pro has a built-in validation feature that will identify missing parameters or tools that aren't properly connected before you run the model. QGIS doesn't have that specific feature, but when you run the model, it will flag missing inputs, outputs, or misconfigured tools.

Here are some best practices to ensure that your model works correctly:

- » **Manually validate parameters.** Before you run the model, double-check that input layers, projections, and attribute fields match the requirements for the tools you're using in the model.
- » **Limit the data.** Run your model with a subset of your data, such as a single neighborhood. Then, using the GIS tools directly and not the automated model, manually calculate results for comparison.
- » **Use descriptive names.** Reduce the likelihood of missteps by renaming your inputs, outputs, and intermediate layers to something meaningful rather than the defaults.

- » **Limit the steps.** For large models, perform step-by-step validation by running one step at a time. Verify the results at each step to quickly pinpoint any issues or errors.
- » **Save as you go:** Save working versions of your model during development so that you can revert to earlier (working!) versions if needed.

Assessing the results

After your model runs without errors, always check the results — *always*. Ask yourself:

- » **Do the results make sense?** For example, if your solar farm site suitability model highlights densely built urban areas as suitable sites, you may have mistakenly included cities rather than excluded them — urban areas typically have limited available land and excessive shading.
- » **Are the outputs too broad or too restrictive?** If your solar siting model shows every hill as suitable or none at all, you probably have an error in your calculations or may need to rethink and refine your requirements.

TIP

Adjust your model's parameters or inputs until the results align with your expectations. Sometimes getting the results to align can take several iterations and a lot of tweaking, but the beauty of modern modeling tools is that they make this process quick and easy.

Gauging whether your model makes sense

Models must make sense. Think beyond the numbers; that is, think about the logic in your model. Are your results consistent with real-world conditions? If you've done solar siting before, you can probably tell right away whether the output from a solar siting model makes sense. If you're new to this type of project, enlist a subject-matter expert to review your model and provide feedback.

REMEMBER

Good documentation and clear explanations come in handy when assessing your model. Fortunately, the modeling tools in both ArcGIS Pro and QGIS allow you to document your process along the way. This documentation helps others understand and validate your process. I can never overemphasize the value of good documentation — not only for others but also for yourself. You'll need it for reference when you have to reuse a model sometime in the future.

Meeting end-user needs

Your model's ultimate success depends on whether it, or the output it produces, meets the needs of the ultimate end users. Keep these tips in mind:

» Always include metadata (data about the data) when providing data with (or as) your final product. (See Chapter 7 for more about metadata.)

» Design output in user-friendly formats, like maps, reports, or dashboards. (See Chapter 17 for guidance on generating output.)

» Use clear symbology and concise legends to improve map readability.

» Deliver results on time and maintain open communication with stakeholders.

TIP

One of the best ways to ensure that you meet the needs of your end users is to keep them involved from the beginning and throughout the process. Their feedback helps to refine your model and ensures that the final products meet their expectations. See Chapter 17 for tips on designing great output.

5
GIS Output and Application

IN THIS PART . . .

Discover what it takes to make effective, accessible output — from printed maps to interactive web maps and apps.

Explore how to automate GIS output using modern techniques.

Find out how GIS transforms the way organizations work and how you can be part of the action.

Chapter 17

Generating Output with GIS

IN THIS CHAPTER

» Creating visually compelling maps

» Selecting colors, fonts, and layouts for effective communication

» Mapping data with classification techniques

» Building interactive maps and apps

» Testing your creation with users

Gone are the days when maps were static images reserved for atlases, wall charts, and gas station road maps. With GIS, map creation has evolved into an art and a science, empowering anyone with the right tools to communicate complex data through visually stunning and highly functional outputs.

In this chapter, you discover how to generate GIS output that goes beyond the basics. Whether you're designing a simple printed map, crafting an interactive web application, or producing noncartographic outputs like reports and alerts, this chapter points you to the tools and techniques to turn raw data into polished results. Along the way, you find out how to design for usability, connect with your audience, and keep your outputs engaging and effective.

This chapter is your guide to confidence — not just as a mapmaker, but as a storyteller who uses GIS to inform, persuade, and inspire.

Exploring Cartographic Design

Creating a map isn't just about plotting data or slapping a north arrow and scale bar on your layout. You're crafting a visual story that communicates clearly and effectively. Whether you're creating a traditional printed map or an interactive web map, good cartographic design is essential.

Understanding key design principles and techniques is the foundation for effective map design, ensuring that your audience can easily grasp your map's meaning.

Reviewing the design process

Great maps don't just happen; they result from careful design. This section offers a cheat sheet to refer to and to jump-start you on map design.

Your design should begin with the two most important things to understand: your audience and the purpose of your map. Start by asking yourself these questions:

» **Who will use this map?** Determine whether your map's users will be technical people, elected officials, researchers, local residents, or others.

» **What decisions or actions should it inspire?** Decide whether your map will be an informational tool, a decision-making visualization, a call to action, or a combination of these.

» **Where will it be viewed?** Design decisions are affected by whether your map will be a poster, printed in a book, displayed on a phone, or projected on a screen.

After you have addressed those questions and have a clear understanding of your map's purpose, follow these steps:

1. **Make a plan by sketching the elements you need and choosing the type of map you want.**

    Decide which elements to include, such as a title, legend, and scale bar. Your map type may be thematic, reference, or topographic; see Chapter 2 for more on map types.

2. **Check your data to ensure that they are accurate, relevant, and properly formatted.**

    If you need to gather more data, check out Chapter 7 for some tips.

3. **Create a first draft of your map and then refine it by adjusting visual elements for clarity and balance.**

   See the later sections in this chapter for help with refining your map.

4. **Review and test.**

   Have others review your map to catch potential errors, typos, and comprehension. Gather feedback to make sure that your design is readable, usable, and makes sense. The last section in this chapter gives you more details on user testing and feedback.

Your map isn't just for you; it's for your audience, too. Think of it as a conversation: if your audience doesn't understand what you're saying, you're not really communicating. Take the time to anticipate their questions and design your map to answer them clearly and intuitively.

Beware of the "curse of expertise"! When you know so much about your subject, it's easy to forget what your audience doesn't know. You might use acronyms that only you understand, cram your map with unnecessary detail, or assume that your audience knows more than they do. Don't fall into the expertise trap. Keep your map simple, clear, and accessible.

Understanding color theory

If cartography had a superpower, it would be color. With color, you can highlight trends, group data, and draw attention to important elements. Colors can also evoke emotions, so choosing the right colors to use in your map is almost as important as using the right data. Bad color choices can confuse readers or, even worse, misrepresent the data.

Color theory can be a book all on its own. In fact, there are a lot of great ones out there, including *Color Theory For Dummies*, by Eric Hibit (Wiley). I don't have room for an entire book within a book, but here are some best practices to get you started and help you use colors effectively:

» **Choose colors that reflect the data, such as**
  - **Sequential data:** Use gradients (light to dark) where darker colors highlight higher values, such as population density.
  - **Diverging data:** For data that has a neutral midpoint, like temperature anomalies, use two contrasting colors with a neutral tone for the midpoint. For example, try red to white to blue or orange to pale yellow to purple for clear contrast.

- **Qualitative data:** Use distinct colors for categories like land use or ethnicity to make them easy to differentiate.

» **Stick to color conventions.** Using conventions makes your maps intuitive. For example, use blue for water, green for vegetation, and red for danger. Be mindful that conventions can vary by culture or context, so keep that in mind when working with colors.

» **Consider accessibility.** Use palettes that work for people with decreased color vision by ensuring a strong contrast between adjacent colors, making them easier to differentiate. See the later section, "Designing for accessibility," for more on this topic.

» **Take advantage of tools for easy color selection.** Some of the more popular online tools include:

- **ColorBrewer (www.colorbrewer2.org):** Provides palettes specific to the type of data (sequential, diverging, qualitative), including options for accessibility for people with color deficiency, making your map printer-friendly, and photocopy safe.

- **MapColPal (www.mapcolpal.org):** Helps you work with color palettes for thematic maps. With this tool, you can generate random colors or load your own and then see how your color choices appear on a live map.

- **Adobe Color (www.color.adobe.com):** Lets you work with and learn the color wheel. With this popular graphic design tool, you can create themes, check for accessibility, and change the color harmony by choosing to include monochromatic, complementary, compound, and analogous colors.

TIP

Start with ColorBrewer for easy palette selection and accessibility options. Then move on to MapColPal to create custom color palettes. Both tools are map centric, so they're perfect for choosing your GIS color palettes and testing accessibility. For even more control over your color choices, explore the Adobe Color tool.

WARNING

Never use your organization's branding colors to highlight negative data, like high crime areas or hazards. You don't want your organization's colors to be associated with bad things!

Choosing appropriate fonts

Fonts, like colors, give your map personality and enhance its readability. A well-chosen font complements your map's design and ensures the legibility of labels and titles. To understand how fonts can affect the overall design of your map, take a look at Figure 17-1. The screenshot on top shows how the California online wildfire-incident map was created. It uses a clean, easy-to-read, and professional looking sans serif font. Next, note the version that uses the comically fun font,

Comic Sans, below it. Can you really take that bottom version of the map seriously? I think not.

A ton of fonts are available, so I know it can be overwhelming. To help you decide what to use, follow these guidelines:

» **Prioritize clarity over trendy.** Although trendy or "fun" fonts may be tempting, resist the urge to use them for your map (refer to Figure 17-1). Save those for your book club flyer or artsy design challenges.

You have two basic font types from which to choose:

- **Sans serif fonts** work great for map labels because they're clean and easy to read at small sizes. Examples include Arial, Open Sans, or (one of my personal favorites) Century Gothic.

- **Serif fonts** add a touch of elegance but are better suited for titles or printed maps for which readability is less constrained. Examples include Times New Roman, Courier, and (another personal favorite) Georgia.

» **Use hierarchy to guide the reader's eye.** Use large, bold fonts for titles and small, light fonts for labels and legends. Avoid overloading your map with too many font sizes; stay with consistent sizes to avoid distractions.

» **Limit font types to avoid a cluttered look.** Don't go overboard by using a variety of different fonts. Use one or two types, such as Georgia for titles and Arial for labels. You can vary weights or styles (bold, italics) as needed.

TIP

Print a sample map and view it from a typical reading distance to ensure that all labels are legible. If you're making a digital map, be sure to check its readability on various devices, including monitors, tablets, and smartphones.

REMEMBER

A font can make or break the tone of your map, so choose wisely. No one wants to see Comic Sans on a wildfire outbreak map!

Designing for accessibility

A well-designed map is meaningless if your audience can't understand it. Accessibility ensures that your maps are understandable to everyone, including people with visual or cognitive challenges. Make your maps more inclusive by doing the following:

» **Use palettes that work for everyone:** Choose color palettes that differentiate colors using both hue and brightness. Tools like ColorBrewer can help you find options that are helpful to people with color challenges. (See the section "Understanding color theory," earlier in this chapter).

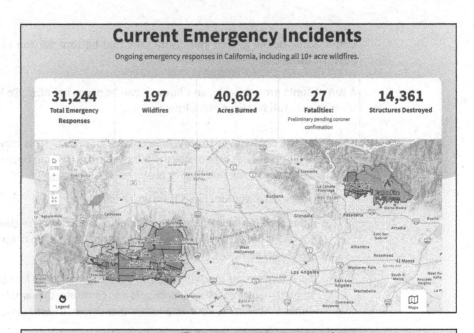

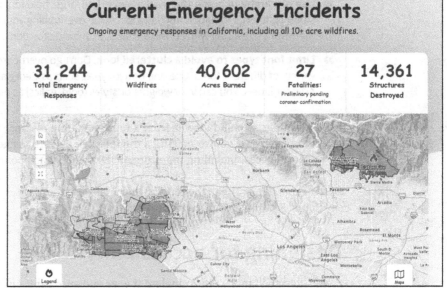

FIGURE 17-1: The 2025 CAL FIRE wildfire map with good (top) and poor (bottom) font choices.

www.fire.ca.gov/incidents (Retrieved January 19, 2025)

» **Make text readable:** Check for sufficient contrast between the text and background using online tools like Adobe Color, WebAIM's Contrast Checker, or Coolors. A contrast ratio of at least 4.5:1 is recommended for small text or fine lines. Pair this approach with readable font sizes and avoid overly decorative fonts.

- » **Employ a scalable design:** Design digital maps to be legible across various devices and resolutions, from larger monitors to smartphones.

- » **Provide alternatives:** Use patterns, textures, or shapes to differentiate your data. For example, instead of using yellow, green, and red circles, choose a yellow circle, a green square, and a red triangle. This method is helpful for audiences with color-vision challenges as well as for black-and-white printing.

Colors convey information, but they're not the only option. Make your maps accessible by using saturation, shape, texture, and text to differentiate data.

Test your map's accessibility using simulator tools for color-vision issues. Although you can find a variety of online tools, both ArcGIS Pro and QGIS include simulators to make testing easy. In ArcGIS Pro, you'll find the simulator on the View tab. In QGIS, it's in Preview mode.

Mapping Data

GIS is all about mapping data. Whether you're visualizing trends, highlighting patterns, or presenting raw facts, the way you map your data is critical for how your audience understands it. But you're not just throwing data on a map; you're telling a story. In this section, you explore the essentials of mapping qualitative and quantitative data and discover how to make complex information clear and compelling using classification techniques.

Mapping qualitative data

Many times, you'll find yourself mapping *qualitative* (descriptive or nominal) data, such as land-use types, political boundaries, or road classifications. Creating these types of maps often involves assigning unique symbols, colors, or patterns to represent each category. For example, you might assign colors such as green to forests, yellow to residential areas, and gray to industrial zones.

Most GIS software displays polygons in a default solid color when you add a polygon layer to your map. The color choice may seem arbitrary at first, but think of it as a blank canvas ready to be customized with colors that best represent your data.

Driving your visual color palette choices are two common attribute types:

- » **Descriptive attributes,** like land-use type or road type
- » **Identifiers,** such as country names or state codes

CHAPTER 17 **Generating Output with GIS** 275

REMEMBER
Not all maps will be in color. For black-and-white printing or accessibility purposes, consider using textures or patterns to distinguish categories. (Refer to the earlier section, "Designing for accessibility," for some tips).

Mapping quantitative data

When you work with numbers, statistics, or other measurements, maps are powerful tools for visualizing trends. How you display that data is just as important as the data itself. Here are three strategies for effective mapping:

» **Show raw counts:** If your goal is to show raw data (like population), map the numbers as is. Be careful, though: Large numbers don't always tell the full story. Imagine looking at the number of newspaper subscribers in New York City — a population of millions — versus Santa Fe, New Mexico, a city with fewer than 100,000 people. This type of map would simply reflect the area's population. Normalizing the data, explained in the next bullet, would make it more meaningful.

» **Normalize your data:** Raw numbers can be misleading when areas vary in size or population. To make comparisons fairer, normalize your data by converting it into rates, percentages, or densities. For example, instead of total population, map population density (people per square mile) or crime rates per 1,000 residents. This way, you can compare large and small areas more accurately.

» **Use ranking:** Replace raw values with rankings to show relative differences. For example, ranking cities by their average income can make comparisons easier to grasp.

REMEMBER
Consider your audience. Do they need raw numbers or are they better served by percentages, rates, or rankings? Tailor your approach to their needs, and be sure to express the data appropriately.

Creating classes

Imagine trying to map 3,000 U.S. counties, each with its own unique shade or pattern. It would look as though a rainbow had exploded. Users would find it visually overwhelming and impossible to interpret. By grouping similar values into classes, you reduce the visual clutter, simplifying the story your map tells. For example, you can group the 3,000 counties into five classes of population ranges.

When creating classes, you need to make two big decisions:

> **How many classes do you need?**
>
> Fewer classes mean simpler maps but may hide important details. Too many classes can make maps hard to read. A good rule of thumb is to aim for 4 to 7 classes.

> **How should you divide the data into groups?**
>
> The method you choose depends on your data and what you're trying to show. Here's a guide to common methods:
>
> - **Equal interval:** This method divides data into equal-sized ranges; for example, 1–100, 101–200, 201–300, and so on. The equal-interval method is simple but can hide patterns if your data isn't evenly distributed. This method works best when you have data ranges that are fairly well-known, such as percentages or temperatures.
>
> - **Defined or fixed interval:** You set the range size, and the software calculates the number of classes. This method works great if you have meaningful ranges in mind; for example, test scores grouped into intervals of 20 points become 1–20, 21–40, 41–60, and so on.
>
> - **Quantiles:** With this approach, each class has the same number of features. This method is great for evenly distributing data visually, but it can distort values if you have outliers in your data. To address this issue, try increasing the number of classes.
>
> - **Natural breaks:** This method is driven by data and finds clusters and gaps in your values. It's a great way to highlight natural patterns in your data. This method is also my personal favorite because it's more likely to represent how the data should logically be grouped. For example, you can use natural breaks to group housing prices to show clear divisions between low, medium, and high costs.
>
> - **Geometric intervals:** This method groups data by creating class ranges that increase in size at a consistent rate, such as by doubling or tripling. It's ideal for datasets that grow quickly or unevenly, such as population growth rates, for which small values are common but large values stand out. (In QGIS, this method is called logarithmic scale.)
>
> - **Standard deviations:** This method is rooted in statistical analysis, dividing data based on their distance from the mean (average). It works best if your data are distributed in a bell shape (that is, normally distributed), like showing temperature anomalies above or below average.

Understanding the different classification types is one thing, but how do you know which one to pick? It comes down to two simple questions: What story do you want to tell, and are your data evenly distributed, clustered, or skewed? For example:

» If you want to highlight natural clusters, use natural breaks.

» If you're comparing values relative to an average, go with standard deviations.

» If you're mapping percentages and temperatures, equal interval might work best.

The right classification method isn't always the most scientific approach; instead, it's the one that makes your data easy to understand and tells the story you want to share. Just be sure not to distort the data in a way that misleads your audience.

Experiment with different classification methods in your GIS software to see which one works best for your data.

Stepping up your classification game

Basic classification methods like natural breaks and defined intervals work well for many maps, but you can take your maps to a higher level if you want. Advanced techniques can provide an extra layer of insight, especially when you're working with complex datasets. Here's how to step up your classification game:

» **Custom breaks** are for those times when the standard classification methods don't quite fit your data. Instead of relying on the software to decide where the breaks should be, you can set your own class boundaries based on specific thresholds or expert advice.

Use custom breaks when you need to highlight key data ranges or when your analysis relies on specific benchmarks. For example, if you're mapping flood-risk areas, historical data may show that regions with more than ten inches of rainfall in 24 hours are high risk. You can use custom breaks to set thresholds based on that historic risk level.

» **Clustering** groups your data based on similarities. Modern GIS tools use machine-learning tools to analyze patterns and suggest meaningful groupings that you might miss with manual classifications. Clustering is perfect for large, diverse datasets with no obvious breaks or patterns. For example, if you're analyzing income data across a country, clustering tools can group areas with similar economic characteristics to reveal trends.

» **Binning and hexbin mapping** simplify dense-point datasets by aggregating points into a square or hexagonal grid. I like hexagon bins because they provide more uniform spatial coverage and fit together seamlessly without gaps. For example, if you're mapping five years of crime incidents in a city, you can aggregate the point locations into hexagonal bins and classify them to show crime hotspots. This approach makes patterns easier to spot without overwhelming your audience.

» **Dynamic and real-time classification** put your audience in control, letting them explore data interactively. Many platforms, like ArcGIS Dashboards or Mapbox, let you use sliders, filters, or custom parameters to update map classifications in real time. For example, you can create a dashboard showing real-time air quality by city, allowing users to filter by pollutant type or time range.

Advanced techniques are powerful, but they also add complexity. Start simple, and move to advanced methods when your data or your audience demands it.

Laying Out Your Map

Laying out your map is a lot like arranging furniture: You need functionality, balance, and just a touch of flair. A skilled cartographer, a graphic designer, or even an interior decorator can create some amazing designs, but you don't have to be one to make a good map. Follow the tips in this section, and you'll be turning your *okay* maps into *really good* ones — maybe even better than that *really* good jerky you bought on your last road trip!

Using essential map elements

Map elements are all the pieces it takes to build a map. They're the parts necessary for reading and understanding your map's story. Here's a quick overview of them (see Figure 17-2):

» **Title:** The title sets the stage by letting readers know what the map is about.

» **Legend:** The legend is your map's decoder ring. It explains the symbols, colors, and classes.

» **Scale bar:** This element provides context for size and distance, helping users understand the map's level of generalization in relation to a location on the Earth itself.

» **North arrow:** Along with the scale bar, the north arrow allows readers to understand the orientation of your map. This symbol is often considered optional when the fact that "up is north" on your map is obvious. However,

CHAPTER 17 **Generating Output with GIS** 279

many GIS folks will argue that it's a required element for every map. Just pose the question to #GISchat friends on social media and you'll see what I mean.

» **Credits:** Give credit where credit is due and always list your sources. And hey, here's where you can take credit for your hard work, too. Nothing wrong with that!

» **Mapped areas:** These areas are the star of the show, and your data takes center stage! Some maps may include an inset map to provide a reference location for the main map.

» **Graticule:** A *graticule* is a grid of longitude and latitude lines, making it useful for a map that requires precise geographic reference. Often used to show scale and position of mapped objects, a graticule is optional, but it makes your map look polished and professional.

» **Borders:** Borders around the map area are sometimes called *neat lines* because they give it an orderly and finished look.

» **Symbols:** Symbols represent real-world features, and no map is complete without them.

» **Labels:** Labels provide names for places on your map so that users can orient themselves. Use bold or slightly larger font size to call attention to features you want to highlight.

REMEMBER

Map element standards evolve over time. What people consider to be essential today may change tomorrow, so keep an eye on current practices.

Factoring in graphic design

A well-designed map is like eye candy: Everything fits together in a nice and sweet way. Good design can turn complex data into an understandable story, whereas bad design can leave your audience lost and confused. The best way to get started designing your map is to think like a graphic designer. I'm not saying that you need to *be* a graphic designer; just think like one. I cover some aspects of thinking like a graphic designer earlier in this chapter, in "Exploring Cartographic Design," which tells you about choosing fonts and using color. If you understand those concepts, you're partway there already! To ensure that your map is designed in a way that tells the story you want to tell, ask yourself these questions:

» **Is it balanced?** Take a look at your map and note where your eyes go when you first look at it. Check whether your scale bar is large and disproportionate to the rest of the elements, or whether one side looks heavy or cluttered. If so, move elements around or resize them to create harmony.

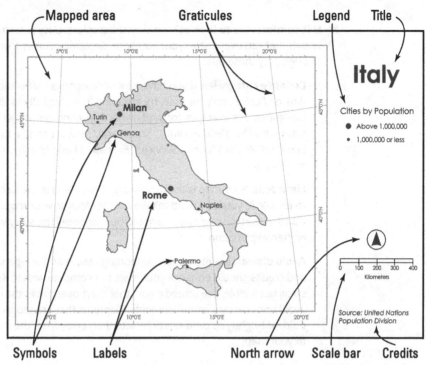

FIGURE 17-2: The anatomy of a map.

» **Can readers see the text?** Fonts should be legible from the distance you expect your audience to view the map. For example, if you're creating a poster, print it at full size, hang it up, and step back to see how it reads. If you're designing a web map, be sure to check it on different screen sizes.

» **Do the features pop?** Use color contrast to make important features stand out. Darker, bold (saturated), and warm (red/yellow undertones) colors tend to stand out, whereas lighter, pale, and cool (blue undertones) colors tend to work well for supporting elements and background. (See the earlier section, "Understanding color theory," for more color tips).

» **Does your map guide the reader's focus?** Use color contrast (see the previous bullet) and visual hierarchy (like different line thicknesses) to emphasize key features and guide the focus of your map. A good map should be able to guide a reader through the story, beginning with the title and then drawing the eyes to the various elements that tell the story (the map, the legend, and maybe some other visuals).

CHAPTER 17 **Generating Output with GIS** 281

>> **Is it comfortable to look at?** Avoid clashing colors, busy patterns, or dizzying visuals that can overwhelm your audience. Removing clutter from your map can help. Try these tips:

- **Collapse categories:** If you have a lot of categories that create an abundance of colors on your map, try grouping them logically to reduce the categories. For example, for land ownership, consider using broad categories like "Federal Land," "State Land," "Tribal Land," and "Private Land" rather than "Forest Service," "Bureau of Land Management," and so on.

- **Limit text:** Your map is the main story. If you need to add a bunch of text to explain it, you may need to take another look at your map. A good visual — that is, your map — should be able to tell the story with limited written explanation.

- **Avoid distracting elements:** Supporting elements like legends, scale bars, and credits should enhance your map, not compete with it. Keep them small but legible, use muted colors that don't overpower the map, and avoid unnecessary embellishments. Elements that are too large, overly bold, or brightly colored can draw attention away from the main story of your map.

The more maps you make, the more you'll discover what works and what doesn't work when it comes to cartographic design. Not everyone can design truly amazing maps that are suitable for framing and hanging, but we can all make *good* maps.

TIP

The best way to learn good design is to look at a lot of maps. Study maps you admire and note what makes them good. Look at bad maps, too. You can often learn good design from bad design because it teaches you what to avoid.

Optimizing maps for different media

Your map may look amazing on your computer screen, but will it look just as amazing on a smartphone, printout, or presentation slide? Optimizing your map for different media ensures that it looks great no matter how or where it's viewed. Here's how to optimize for these types of media:

>> **Digital Maps:** Use interactive features like zoom, pop-ups, and dynamic legends to enhance usability. Also, choose colors that display consistently on a variety of devices, including older ones.

- **Printed Maps:** Use high-resolution (300 DPI or higher) for clear, professional-quality prints and stick to CMYK-friendly colors to avoid dull or distorted prints.

- **Presentations:** Simplify your design by removing unnecessary details and enlarging fonts to ensure readability. Use contrasting colors to make elements stand out, even in brightly lit rooms. Set your layout to match the slide size (16:9 widescreen or 4:3 standard) so that it fits perfectly.

TIP

Always test your map in its intended format before sharing it. What looks perfect on your screen may need tweaking for print or web use.

Developing Interactive Maps and Apps

Gone are the days when maps were just static images pinned to a wall. Today, interactive maps let you zoom, click, and even ask questions, transforming how people interact with spatial data.

Unlike desktop GIS software (such as ArcGIS Pro and QGIS), which is great for handling large datasets, automating workflows, and performing advanced spatial analysis, web mapping platforms are designed primarily for interactivity, collaboration, and accessibility. Although some web tools offer basic spatial analysis, their real strength lies in making data easy to explore and share.

This section walks you through the tools and techniques you need to get started on creating a simple web map or a full-fledged app.

Choosing tools and platforms for web mapping

The first step in building an interactive web map is to pick the right platform. Luckily, plenty of tools are designed to help you — no programming degree required! Here are a few popular ones:

- **ArcGIS Online:** This cloud-based mapping tool by Esri makes it easy to upload data and create interactive web maps with built-in templates and widgets.

- **Felt:** Felt is a collaborative mapping platform that's great for creating and sharing simple, beautiful maps quickly. Its intuitive interface makes it ideal for team projects.

CHAPTER 17 **Generating Output with GIS** 283

- **QGIS2Web:** If you're a QGIS fan, this plug-in lets you export your maps directly to interactive open-source tools like Leaflet, Mapbox, or OpenLayers.
- **Mapbox:** Mapbox is a flexible tool for creating custom, visually striking maps. It's especially popular for maps that need heavy customization.
- **Kepler.gl:** An open-source tool designed for geospatial data analysis, this tool is great for visualizing large-scale geospatial data with beautiful, interactive graphics.
- **Google Maps Platform:** Ideal for integrating maps into websites or apps, Google Maps Platform offers strong tools for geocoding and routing.

In case you're not quite sure where to start, each of these tools offers free tutorials. You can try them to find out which one works best for you.

Enhancing maps with interactive features

Adding interactivity turns a basic map into an engaging experience. With interactive maps, you can enable users to find the urgent care center that's closest to their home, research the demographics of an area, or discover how wetlands have changed over time. Here are some elements you can use to add interactivity to your maps:

- **Pop-ups:** You can use pop-ups to show more information about a feature when users click it. For example, clicking a city might display a pop-up containing its population, average income, or historical facts. You can even add style to your pop-ups by adding photos, charts, or graphics.
- **Filters and layers:** Let users turn map layers on and off to see the features that interest them or filter data to focus on specific details.
- **Zoom and pan:** Add intuitive navigation so that users can explore data across different scales.
- **Time sliders:** Display data changes over time, such as by tracking weather patterns or historical migration trends.

Don't overwhelm your audience with too many features at one time. Stick with what's relevant. Keep your map clean, informative, and intuitive!

Creating data dashboards

Data dashboards combine maps, charts, lists, and other visuals to tell a quick story. They're perfect for sharing real-time updates, tracking performance metrics, or summarizing trends. Here are some of the more popular dashboard platforms:

- **ArcGIS Dashboards:** The dashboard platform from Esri makes it simple to combine maps with real-time graphs, gauges, and lists.
- **Grafana:** This open-source dashboard tool is excellent for real-time data monitoring, with options to include geospatial visualizations using the geomap panel.
- **Power BI with Map Visuals:** This dashboard tool from Microsoft integrates mapping plug-ins like ArcGIS for Power BI, giving your dashboards geographical context.
- **Tableau:** Although not specifically GIS-focused, Tableau integrates spatial data into polished, interactive dashboards. It's one of the most popular data-visualization platforms.

Dashboards are extremely useful for providing critical information. For example, if you're working in wildfire management, you can create a dashboard to monitor wildfire response. Include a map of active fires, bar charts showing evacuation counts, and a gauge to indicate available fire suppression resources.

Developing GIS-based web applications

If you're ready to go beyond dashboards and simple interactive maps, you can use the following GIS-based web applications to create fully customized tools:

- **ArcGIS Experience Builder:** Build custom apps with a drag-and-drop interface — no programming required. You can quickly add widgets like charts, images, and interactive maps to create tailored experiences. Experience Builder is great for beginners and anyone who wants customization without diving into code.
- **Google AppSheet:** A lightweight, no-code platform that integrates Google Sheets, Excel, and other cloud-based data sources. Although not GIS-specific, Google AppSheet can incorporate maps and location data, making it a beginner-friendly option for simple applications. AppSheet is ideal if you want quick results without the complexity of many GIS-focused platforms.
- **Leaflet and OpenLayers:** If you're comfortable with code, you can use these open-source JavaScript libraries to create lightweight, highly customizable web maps. These tools are great for developers and code-tinkerers who want full creative control.

» **Esri developer APIs:** If you're working with ArcGIS data, the application programming interfaces (APIs) from Esri for JavaScript, Python, and more are a good choice. They give you full control over how your maps look, behave, and interact with other data. If you're an advanced user looking for a deep-dive into GIS-powered applications, these APIs are for you.

If you're new to app development, take it step by step! Start small with tools like Experience Builder or AppSheet to get a feel for app creation. When you're comfortable, explore coding-based tools for more flexibility and customization.

Incorporating user interaction and feedback

You know you're doing something right when your interactive maps and apps truly connect with their intended audience. After all, a map that's easy to use and invites feedback keeps people engaged and coming back for more. Here's how to make sure that your creation resonates:

» **Design for usability:** Keep it simple and intuitive. Features like zoom, filters, and legends should be easy to find and use, so put them where people expect to find them! No one wants to hunt for a Zoom button. Think about how your audience will interact with the map and make their experience effortless.

» **Gather feedback:** Your map is only as good as its usefulness. Invite colleagues — or even friends and family — to test your map and share their thoughts. Sit quietly while they explore, and take notes on their experience. Also, add a Feedback button or form so that users can suggest improvements, report issues, or just share what they love. You may be surprised at the insights they provide!

» **Test, test, test, and test again:** Don't let your incredible creation flop because it doesn't work on a smartphone. Test it on as many devices as possible — desktops, tablets, phones, and even different browsers. Check it using accessibility tools like simulators for people with color challenges to make your map inclusive. (The section "Designing for accessibility," earlier in this chapter, has more details about making your map accessible.) If something doesn't work, fix it before your users spot it. (See the section "Testing and User Feedback," later in this chapter.)

» **Engage your audience:** Got a specific goal for your map? Spell it out! Whether you're promoting community projects or encouraging survey participation, make your call to action clear and inviting. A little guidance goes a long way.

TIP The best interactive maps are always a work in progress. Incorporate feedback early and often, and watch your map evolve into something better than you imagined.

Creating Noncartographic Outputs

Maps may be the first thing that comes to mind when you think about GIS, but noncartographic outputs help you take action. From automating reports to monitoring real-time events, these tools complement and enhance maps, helping you do more than just visualize your data.

Automating reports and data exports

Automation adds an "easy button" to your work. By automating, you can turn raw GIS data into polished, actionable outputs without breaking a sweat. Automation not only saves time but also ensures that your results are consistent, reliable, and ready when you need them. Here are a few types of automation to add to your workflows:

» **Regular exports:** Automate exports to popular formats like Excel, PDF, or CSV. For example, you can export monthly summaries of customer density or store performance to integrate with customer relationship management (CRM) systems.

» **Template-based reports:** Use GIS tools to generate reports with predefined layouts, complete with maps, charts, and tables. For example, a quarterly environmental impact report might feature air-quality graphs alongside annotated maps. You can also automate report generation and distribution, ensuring that stakeholders receive regular updates via email.

» **Real-time data feeds:** Automate dynamic data exports like traffic counts, weather data, or sensor readings. If you're sufficiently tech savvy, you can set up APIs to provide real-time access to your organization's data.

TIP In case you're not a coding guru, ArcGIS Pro's Tasks and QGIS's Model Designer are great for setting up workflows without writing scripts. To get started with automation in GIS, see Chapter 18.

Producing lists and summary statistics

Sometimes simple is better. Lists and summary statistics break down complex datasets into clear, actionable information, making them ideal for reports, presentations, or decision-making. GIS can produce a variety of lists; for example:

» **Attribute-based lists:** Pull records based on specific criteria, like all parcels owned by the city or properties within a flood zone. A list of flood-prone properties can guide insurance companies or emergency responders.

» **Spatially defined lists:** Generate lists of features within a geographic area, such as schools within two miles of a proposed development, or shelters in an evacuation zone. These lists help planners or disaster teams prioritize areas for action.

» **Hierarchical lists:** Sort and group features based on attributes, like population by county or sales by district, which is useful information for resource allocation or market analysis.

With GIS, you can also calculate key metrics from your data, including the following:

» **Counts and totals:** Quick calculations, like how many schools are within a region, or the total forested area in the county, can support budget planning or environmental assessments.

» **Averages and ranges:** Averages or ranges, such as the average household income in the city, or how temperatures vary across a climate zone, can help to inform city planning or climate adaptation efforts.

» **Trends over time:** You can summarize changes in wetland areas, urban growth, or weather patterns across multiple years. For example, spotting patterns in urban growth helps planners make informed decisions about zoning and infrastructure.

TIP

GIS software has built-in tools for calculating statistics and summarizing data. See Chapter 8 if you're working with raster data and Chapter 9 for working with vector-data attributes. For large datasets, use spreadsheets or dashboarding software like Tableau or Power BI to discover trends or generate interactive data visualizations.

Creating systems for real-time monitoring and alerts

Sometimes you need to know what's happening *right now*, and GIS can help you do that. Real-time GIS systems pull data from sensors, GPS devices, or live feeds and display it in easy-to-read formats like dashboards. Tools such as ArcGIS Dashboards or open-source alternatives like Grafana can help you share real-time updates visually. Here are some ways you can use GIS monitoring and alert systems:

- » **Wildfire monitoring** enables you to send evacuation alerts when fire lines approach communities.
- » **Traffic congestion:** You can update congestion maps every minute to help drivers avoid bottlenecks.
- » **Pipeline safety:** You can trigger alerts if pressure drops, signaling a potential leak in the system.

REMEMBER

Real-time systems are only as good as their data. Use reliable sources and test your alerts to ensure that they're accurate and timely.

Testing and User Feedback

Good GIS output isn't just about creating maps and reports, it's also about ensuring that your output is clear, useful, and accessible to your audience. Usability testing and feedback help you fine-tune your maps, reports, and apps so that they work seamlessly for your audience.

When I teach GIS or data visualization workshops, I always start with the importance of knowing your audience. I revisit and emphasize this point throughout the course because it's so critical. So although I mention this idea earlier in the book, it's worth stressing it again: At the start of a project, you should have a good sense of your audience and their needs. Ongoing testing and user feedback help you stay aligned with those needs and guide improvements to ensure your project continues to meet them. This section offers tips on how to conduct usability testing, gather feedback from your audience, and update your GIS project based on the user input.

Conducting usability testing

Before sharing your outputs with the world, test them like you mean it! Usability testing involves observing how real users interact with your creation and identifying any issues. Whether you're working on a printed map or an interactive app, the goal is the same: to ensure that your output works for your audience. Here are some best practices to keep in mind:

> » **Test with your audience in mind:** Choose testers who match your audience — whether they're planners, stakeholders, or constituents.
>
> » **Test for clarity:** Ask users to identify key features or interpret patterns on your map. Do they understand the map without any explanation from you? Are any elements confusing or distracting?
>
> » **Focus on key tasks:** Ask your testers to complete common tasks, like finding a specific location on a map or generating a report.
>
> » **Watch and learn:** Pay attention to where users struggle. Are buttons hard to find? Is the interface too confusing? Take notes on what needs fixing, and fix it.

TIP

Sit quietly while your testers work and resist the urge to explain or help. Observing silently is the best way to get honest feedback.

Gathering feedback from diverse audiences

I had a professor in college who said, "There's no such thing as the *general public* — it's too broad of a term. Focus on specifics about your audience." He was right. Although your audience may be diverse, they won't have one-size-fits-all needs. Your job is to think about who your users are and tailor your outputs to their skill levels and expectations. Then gather meaningful feedback by

> » **Inviting a range of testers:** Include users with different backgrounds, levels of GIS expertise, and accessibility needs aligned with your target audience.
>
> » **Asking the right questions:** Focus on clarity, ease of use, and whether the output meets their needs. For example, you might ask:
>
> - What did you find most helpful?
> - What was confusing or frustrating?
> - What would make this better for you?
>
> » **Providing different ways to give feedback:** Use forms, surveys, or open forums for users to easily share their thoughts. For printed maps, you can include physical forms or host in-person focus groups.

Iterating based on user input

Gathering feedback is just the first step. Acting on that input is what transforms good GIS output into great GIS output. Here's how to continuously improve:

- **Prioritize fixes.** Address the most critical issues first, like improving text legibility on a printed map or fixing a broken link in an app.
- **Keep improving.** Even small tweaks, like changing font sizes, changing color palettes, or adding labels, can make a big difference in usability.
- **Test again.** After making updates, retest with users to ensure that the changes work as intended. For printed maps, this testing may simply involve printing a revised draft to verify improvements. For apps, it may mean another round of usability testing.

Think of your GIS output as a living document, always evolving and improving with user input. This approach doesn't just create better maps; it builds trust and credibility with your audience.

IN THIS CHAPTER

» Exploring how you can use programming languages in GIS

» Applying programming to real-world GIS tasks

» Accessing data with APIs and web services

» Following best practices for GIS automation

Chapter **18**

Automating GIS

Automation is a GIS power tool that helps you streamline repetitive tasks, analyze large datasets, and integrate complex workflows with ease. Even with a basic understanding of programming, you can save time, improve accuracy, and unlock capabilities that traditional GIS interfaces can't match. Whether you're customizing a script for data processing or building an interactive web map, automating GIS opens you to a world of possibilities.

This chapter offers you a glimpse of the common programming languages used in GIS. You also explore simple examples of how automation can enhance your workflows, improve efficiency, and help you work smarter. This short chapter doesn't show you how to program, but it gives you a taste of what's possible and helps you start your journey in the world of GIS automation.

Getting to Know GIS Programming Languages

GIS programming languages come in all shapes and sizes, from powerful, versatile options like Python and JavaScript to specialized tools built just for GIS platforms. These languages let you to automate workflows, tackle big datasets, and create custom tools. Don't worry if heavy programming isn't your thing; many GIS platforms offer lightweight, beginner-friendly scripting options that don't require advanced coding skills.

Here's a closer look at some commonly used languages in GIS:

» **Python:** Python is the Swiss Army knife of GIS programming. It's used for everything from automating geoprocessing workflows in ArcGIS Pro or QGIS to building web apps with geospatial data. Libraries like arcpy (for Esri software) and geopandas (an open source alternative) make it easy to manipulate spatial data with just a few lines of code.

For example, these two lines of Python code automate the task of creating a buffer around roads:

```
Import arcpy
arcpy.Buffer_analysis("Roads", "Buffered_Roads",
    "100 Meters")
```

» **JavaScript:** If web mapping is your goal, JavaScript is the go-to language. JavaScript, along with libraries like Leaflet or Mapbox GL JS, powers interactive online maps and apps. Esri's ArcGIS API for Javascript is another popular option for creating customized web mapping applications.

In this example, JavaScript is used to create a simple map using the Leaflet library:

```
Var map = L.map('map').setView([51.505, -0.09], 13);
L.tileLayer('https://{2}.tile.openstreetmap.org/{z}/{x}/
    {y}.png').addTo(map);
```

» **R:** Although not traditionally associated with GIS, R is a powerful statistical programming language with excellent spatial data support. Packages like sf (for handling spatial data), tmap (for creating maps), and tidycensus (for accessing Census data) make R a favorite for geospatial data analysis.

The following example snippet of R code reads in the roads.shp shapefile and then creates a buffer around the road features:

```
library(sf)
roads <- st_read("roads.shp")
buffer <- st_buffer(roads, dist=100)
```

» **Arcade (Esri):** Built into Esri platforms like ArcGIS Pro and ArcGIS Online, Arcade makes on-the-fly calculations and dynamic labeling accessible without the need for advanced coding skills. However, you do need to grasp basic coding concepts, like writing expressions and referencing attributes.

This example snippet dynamically labels features with formatted population data:

```
"Population: " + Text($feature["Population"], "#,###")
```

» **QGIS Expressions:** Similar to Arcade, QGIS Expressions are great for quick spatial calculations or filtering data directly in the interface. Like Arcade, expressions don't require you to be a programming expert, but a basic understanding of coding syntax and attribute reference is helpful.

For example, this QGIS Expression converts a feature's length from meters to kilometers:

```
$length / 1000
```

» **SQL:** Structured Query Language (SQL) is essential for working with geodatabases or querying spatial data. Both Esri and QGIS support SQL queries for filtering or analyzing data stored in databases.

This example SQL query selects all records from a parcels layer that are zoned residential:

```
SELECT * FROM parcels WHERE zoning = 'Residential';
```

TIP

Start with a programming language that aligns best with your goals. For example, Python is great for automating tasks, JavaScript excels at web mapping, and Arcade or QGIS Expressions are perfect for quick calculations. After you've mastered one, the others will feel much less daunting!

DECODING THE PROGRAMMING JARGON

To help you avoid drowning in the deep end of the programming jargon pool, here's a primer on some key programming terms you'll likely come across in your GIS journey:

Program or script: What's the difference?

You may hear the terms *program* and *script* terms used interchangeably, but they're not exactly the same. A *program* is a full-fledged application that performs complex tasks, often with a user interface and multiple components. For example, ArcGIS Pro and QGIS are programs. A *script* is a smaller piece of code (although it can still be long and complex) that's written to automate a specific task, like geocoding a list of addresses or generating a map. Scripts are faster to write and easier to modify than programs.

(continued)

(continued)

To return to the pool metaphor, a program is like a water park offering a variety of features for an all-day adventure, whereas a script is like a lap pool that's built for focused, efficient use.

Libraries — but where are the books?

In programming, a *library* is a collection of prewritten scripts that help you avoid reinventing the wheel. For example, Python libraries like arcpy (Esri's library for GIS tasks) or geopandas (for geospatial data in Python) provide tools for common tasks like spatial analysis or coordinate conversions. Think of a programming library as a bookshelf full of useful tools: Referencing a function or method from a library in your script is like pulling a book off the shelf and using it rather than writing your own book from scratch.

API or web service: Aren't they the same thing?

Both an application programming interface (API) and a web service let you interact with software or services online, but they're not quite the same:

- **API:** An API allows your script to programmatically interact with a web service. For example, the Google Maps API lets you add maps or routing capabilities to your app. Think of an API as a restaurant menu that shows you what's available and how to ask for it.

- **Web service:** A web service is the actual online functionality or data provided, and you often access it through an API. For example, the OpenStreetMap web service provides map tiles (pre-made raster or vector-based sections of a map that load dynamically as you move around the map) or geospatial data. If the API is the menu, the web service is the chef preparing the meal. Together, they make a seamless dining (or coding) experience.

Making GIS Work Smarter with Scripts

Creating scripts opens a whole new world of possibilities by letting you automate repetitive tasks, extend existing tools, and create custom workflows. Here are some cool things to know about scripts:

>> **Automate tasks:** If you're tired of manually exporting dozens of maps or manually reviewing a large dataset for missing attributes, you can write a script! For instance, you can write a Python script to batch process satellite imagery or validate your spatial data.

» **Analyze and visualize data:** Scripting expands what you can do with GIS analysis by enabling you to handle complex calculations or create custom visualizations. You can use Python to perform statistical analysis on spatial data or JavaScript to build interactive web maps. The possibilities are endless.

» **Connect to online services:** Say you need live weather data or real-time traffic updates for a GIS project. APIs let you integrate external data directly into your project. For example, you can pull weather data from the OpenWeatherMap API and overlay it on a map to see storm patterns. I tell you more about APIs in the next section.

» **Use visual workflows:** In case you're not ready to dive into scripting just yet, tools like ModelBuilder from Esri let you automate tasks through a drag-and-drop interface. ModelBuilder can serve as a stepping stone into scripting by helping you visualize workflows and generate reusable models.

» **Streamline data integration with ETL tools:** Working with spatial data often involves extracting it from different sources, transforming it into a usable format, and loading it into GIS, a process known as extract, transform, load (ETL). Instead of manually converting and cleaning data, you use ETL tools like FME (Feature Manipulation Engine) by Safe Software or FME Workbench by Esri to automate these workflows. For example, an FME workflow can extract CAD data, transform it into a GIS-friendly format, and load it into an Esri geodatabase, all without the need for extensive scripting. ETL tools are especially useful for reducing errors while handling large, complex datasets.

» **Develop web maps and apps:** JavaScript libraries like Leaflet or ArcGIS API from Esri for JavaScript make it easy to build interactive maps that you can share online. Whether it's a public-facing app showing bike-share locations or an internal tool for tracking deliveries, GIS programming brings your ideas to life.

Getting Data through APIs and Web Services

Application programming interfaces (APIs) and web services are your backstage passes to a world of geospatial data. They're your ticket to fetch, update, and integrate data directly into your GIS workflows without the hassle of manual downloads. With just a few lines of code, you can automate data retrieval and keep your projects updated effortlessly. Imagine making daily updates to your county's wildfire watch map using real-time weather data. An API makes such updates possible!

APIs act as intermediaries, allowing your scripts to request data from a server. For example, a weather API can provide real-time temperature and precipitation data for specific locations, which you can integrate into a GIS project with a simple script to create a live weather map.

TIP

To avoid unexpected surprises, always check the API documentation for usage limits.

Here are some useful GIS APIs and example uses for you to explore:

» **Esri REST API:** Access map services, geocoding tools, and spatial data directly from ArcGIS Online.

To build a property assessment tool, you can use REST API from Esri to automate adding map layers or querying spatial data directly in ArcGIS Online.

» **Google Maps API:** Retrieve geocoding, routing, and place information.

To improve delivery times, you can use the Google Maps API to fetch and visualize live traffic conditions directly in your map.

» **OpenStreetMap Overpass API:** Extract specific OSM features like roads, buildings, or parks for custom use.

To support urban planning, you can extract detailed building footprints using the Overpass API to plan more effectively and efficiently.

» **NASA Earthdata API:** Access remote imagery and Earth science data.

To research deforestation, you can use the Earthdata API to automate downloading satellite imagery of deforestation hotspots.

» **OpenWeatherMap API:** Fetch current and forecasted weather data.

To track weather patterns over time, you can automate daily data pulls with the OpenWeatherMap API, enabling you to create a dynamic, up-to-date GIS dashboard. (Check out the sample R script in Chapter 7 that uses this API).

Understanding Best Practices

Automation can save you time and effort, but as with any tool, it's most effective when used wisely. A poorly written or unorganized script can quickly turn into a headache with even more problems — and believe me, I've had my fair share! By following some simple programming best practices, you can make your scripts more efficient, adaptable, and maintainable — all without needing aspirin.

Whether you're reusing code, handling errors, optimizing performance, or documenting your work, the tips in this section will help you get the most out of your automation efforts.

Reusing the code

Why reinvent the wheel when you can reuse it? Writing modular, reusable scripts not only saves time but also creates building blocks for larger workflows. By breaking tasks into smaller components and designing scripts with flexibility in mind, you set yourself up for long-term success.

Here's how to make your code adaptable and reusable:

» **Use a modular design:** Start small, and focus on automating one task at time. You can combine modular scripts into more complex workflows later. For example, a function for creating a buffer around a feature can be reused across multiple projects. (See Chapter 9 for more about creating buffers.)

» **Take advantage of parameterization:** This fancy word simply refers to writing scripts that accept inputs (parameters) so that you can adapt them to different datasets without rewriting the code. For instance, instead of *hardcoding* file paths (writing the file path directly into your script), you can create code that prompts the user to select the input file at runtime. This approach makes your script more flexible and reusable.

TIP

To save yourself some time and effort on future projects, create a personal library of scripts and code snippets for common GIS tasks. These reusable items can serve as building blocks for larger scripts and programs.

Handling errors

Errors in scripts and programs happen. It's simply a fact of life. The best way to deal with errors is to anticipate them. Plan for issues like missing files, invalid data formats, or incorrect input, and handle them within your code with clear error messages or fallback workflows.

For example, add an error message if a file is not found:

```
If not os.path.exists(input_file):
        print("Error: Input file not found")
```

You can also build fallback workflows to keep your process running. For instance, if a required dataset is missing, your script can notify the user while skipping to the next available task. This fallback approach keeps automation on track without the need for constant oversight.

Optimizing performance

Automation is about working smarter, not harder — and that idea applies to your code's performance, too. Efficient scripts run faster, handle larger datasets, and make better use of your system's resources. Consider these strategies to avoid bottlenecks and crashes:

- » **Process in chunks:** Break large datasets into smaller parts to avoid overwhelming your system's memory. For example, if you're processing a massive raster file, handle it tile by tile instead of loading the entire file all at once. (A tile is a pre-made section of a larger raster file, often used to efficiently display large datasets).

- » **Leverage built-in tools:** Use built-in GIS functions where possible. Python libraries like arcpy and geopandas include built-in tools that are optimized for performance, often outperforming the use of custom-coded loops.

TIP

To ensure that your scripts run efficiently, test them on smaller datasets before scaling up to the full dataset.

Documenting the process

Solid documentation turns good code into great tools. Whether you're revisiting your work months later or sharing scripts with a colleague, well-written documentation ensures that your code is understandable and useful.

Here are some tips for how to document effectively:

- » **Comment your code:** Add clear, concise comments to explain what your code does. Commenting makes it easier for others (or future you) to understand your logic. In Python, you have two main ways to add comments:
  - Use # for inline and single-line comments to explain specific parts of your code.
  - Use triple quotes """ (also called docstrings) inside functions to provide structured documentation.

For example, the following script shows best practices for using these comment types in a Python script:

```python
# This function creates a buffer around an input shapefile
import arcpy

def buffer_shapefile(input_shp, output_shp,
    buffer_distance):
"""Creates a buffer around features in a shapefile."""

# Run the Buffer tool with the specified distance
arcpy.Buffer_analysis(input_shp, output_shp,
    buffer_distance)

# Print confirmation message
print(f"Created a {buffer_distance} buffer around {input_
    shp} and saved it as {output_shp}.")

# Example usage:
# buffer_shapefile("roads.shp", "roads_buffer.shp",
    "1000 Meters")
```

» **Include plain documentation:** Create a simple text file to explain your script's purpose, its required inputs, and how to run it. For instance, you might include a short readme.txt documentation file with your script.

» **Maintain version control:** Use tools like Git, GitHub, and Bitbucket to track changes and collaborate with others.

Good documentation doesn't have to be fancy — a few notes in a plain text file can make all the difference!

IN THIS CHAPTER

» Exploring how organizational structures affect GIS success

» Adapting GIS for diverse organizational needs

» Building your career and professional network in GIS

Chapter **19**
GIS in Organizations and Career Development

A wide range of organizations — such as government agencies, research institutions, and retail chains — uses GIS to transform the way they operate, make decisions, and meet their goals. But getting the most out of GIS takes more than just installing software and hiring a GIS analyst. You also need clear communication, collaboration, and thoughtful planning to align GIS with your organization's structure and objectives.

More than just hardware and software for making maps, GIS is a tool that can transform how organizations operate, make decisions, and meet their goals. But successfully integrating GIS into your organization requires a much greater effort than simply installing software and hiring a GIS analyst. You need clear communication, collaboration, and careful planning to align GIS with your organization's structure and objectives.

In this chapter, you explore how GIS shapes organizations, how you can tackle common challenges during implementation, and how to plan for long-term success. You also find practical advice for building your career in GIS, including developing key skills and growing your professional network.

Whether you're working to bring GIS into your organization or carving out your own career path, this chapter's ideas can help you get there.

Transforming Organizations with GIS

Far from just mapping locations, GIS also reshapes how organizations operate and make decisions. Integrating GIS into your organization enables you to gain insights that can improve efficiency, collaboration, and planning — a true trifecta! To achieve those improvements, you need to know the roles that GIS plays within your organization and tailor your GIS program to meet your organization's unique needs.

Exploring GIS roles and interactions

GIS transforms both the way an organization works and how teams work together. Internally, you collaborate with GIS operators, managers, and other personnel to drive projects forward. Externally, you tap into the larger GIS community by gathering data, delivering results, purchasing equipment, and staying up to date with training. Every team member, whether internal or external, plays a role in making GIS a success.

Internal teams

Here's a breakdown of the key roles within internal teams and how those roles interact:

- **GIS analysts:** These are the problem solvers who perform spatial analysis, create maps, and build models that inform critical decisions. They often work across departments, collecting, sharing, and translating data into useful information.

- **GIS technicians:** These specialists keep everything GIS-related running smoothly. They maintain spatial data, perform quality control, and ensure the accuracy of the organization's GIS datasets. Their work ensures the reliability of GIS products.

- **Application developers:** These are the builders who create geospatial experiences. They design custom tools, applications, and user interfaces tailored to the organization's needs. They enhance the usability of GIS by integrating advanced functionality.

- **Decision makers:** The decision makers are the GIS team's champions. Although not directly involved in GIS operations, these organization leaders rely on GIS outputs to make data-driven decisions. Empower them with intuitive dashboards or interactive geospatial apps, and they'll become your GIS champion.

External partners

Your external partners serve your organization in these important ways:

» **Data providers:** From public sources like government data portals to private companies offering value-added datasets, data providers supply the fuel for GIS operations.

» **Technology vendors:** Modern GIS software companies supply more than just stand-alone desktop apps. The emphasis now is on offering reliable direct connections to online data portals and cloud-based solutions, such as ArcGIS Online or Mapbox, thereby making GIS scalable and accessible.

» **Consultants and freelancers:** For short-term projects or specialized needs, organizations collaborate with external GIS professionals to enhance capacity and expertise.

Collaboration is one of the superpowers of GIS. Understanding the various team roles both inside and outside the organization is one part of this collaboration. Establishing clear communication, encouraging data sharing, and coming together on project goals is the other.

Adapting GIS to organizational needs

No two organizations use GIS the same way. Whether you're part of a large corporation, a mid-size government agency, or a small nonprofit, you can adapt GIS to fit the needs of your organization. (See the nearby sidebar, Adapting GIS for small organizations," for tips on making GIS work with limited resources). To successfully integrate GIS, it's important to tailor it to align with your organization's specific goals and workflows. Here are some best practices to follow:

» **Understand workflows:** Check out how the data flows within your organization and identify the areas where GIS can add value, such as by automating routine tasks or enabling spatial analysis.

» **Focus on the audience:** Keep your audience's needs front and center. Whether you're working with an environmental nonprofit prioritizing habitat maps or a logistics company optimizing routes, be sure to align GIS products with their goals.

» **Be aware of ethical considerations:** With the rise of open data and real-time monitoring, privacy and ethical challenges concern every organization. Ensure that your GIS policies and practices comply with legal standards and promote transparency.

» **Stay abreast of scalability and innovation:** If given a choice, always opt for technologies that can grow with your organization. Cloud-based GIS platforms, real-time dashboards, and integrations with tools like IoT (Internet of Things) sensors or machine-learning models keep your GIS ready for the future.

Tailoring GIS to your organization's needs isn't a solo job. Engage key users early to help them understand workflows and goals. Doing so enables your system to deliver value and creates a sense of ownership among all stakeholders.

ADAPTING GIS FOR SMALL ORGANIZATIONS

Even small teams can leverage GIS effectively. Although small organizations often face challenges like limited budgets, fewer staff, or restricted access to resources, GIS isn't out of reach. With proper planning, it can be a real game-changer for small businesses, governments, and nonprofits. Here's how to make it work:

- **Start with free or low-cost tools.** Explore options like QGIS or cloud-based GIS platforms that reduce up-front costs. Many tools offer powerful capabilities without requiring heavy IT infrastructure. Registered nonprofits can also apply for special pricing from Esri.

- **Collaborate with partners.** Universities, nonprofits, and even local government agencies often provide access to data, training, or expertise that can support your GIS efforts. If you're a small government agency, check with your state or regional GIS office to see whether they can extend their enterprise GIS license to smaller agencies in their jurisdiction.

- **Focus on small wins.** To gain management buy-in, start with manageable projects that show immediate value, like creating maps to support grant applications or improving data visualization for reports.

- **Leverage open data.** Publicly available datasets from government portals or open-source communities get you up and running quickly without the need for costly data subscriptions. (See Chapter 21 for places to start).

GIS doesn't have to break the bank to deliver big results. With the right tools and partnerships, even small organizations can achieve great things.

Designing and Implementing GIS in Your Organization

Every organization's culture, structure, goals, and workflows are unique. All these aspects of an organization greatly influence how or whether it can effectively use GIS (or any technology, for that matter) and whether it can easily integrate that technology into existing workflows. To overcome integration challenges and plan for long-term growth, you need a thorough understanding of your organization's specific goals and needs.

Aligning GIS with organizational goals

You can't integrate GIS if you keep it in a silo. Integration means ensuring that the entire organization benefits from GIS. To break free of the "silo effect," make sure that GIS workflows align with your organization's objectives. Here are some practical steps to get you started:

- » **Understand organizational structure.** Grab a copy of the current org chart to see where GIS fits and who will benefit from it. Map out how decisions flow, who holds responsibilities, and where GIS can provide the most value.

- » **Clearly define goals.** Establish specific, measurable objectives for GIS implementation. Are you improving decision-making? Streamlining workflows? Expanding your data analysis capabilities? Tailor GIS to meet these targets.

- » **Promote data sharing.** Encourage data sharing across departments to break down silos. This ensures everyone has access to the same data sources for consistent analyses and reporting.

- » **Prioritize training.** If you're leading a team, make sure they understand both geographic principles and GIS technology. Equip staff with the skills needed to leverage GIS effectively, from basic mapping to advanced analysis.

REMEMBER

Involve all stakeholders early and often, and make sure that your GIS aligns with both organizational goals and practical needs by gathering input from decision makers, IT teams, and end users. Along the way, you'll discover that the more often you keep internal teams informed, the more likely they are to work *with* you rather than *against* you.

CHAPTER 19 **GIS in Organizations and Career Development** 307

Overcoming integration challenges

Bringing any new technology into an organization changes the way people work, affecting everything from workflows to team culture. Change always comes with a fair share of challenges. Not everyone will embrace the change; you can expect some resistance, confusion, and distrust. Here are some tips for meeting some of these challenges head on:

>> **Manage organizational resistance to change.** Resistance often comes down to entrenched processes. Departments or teams may hesitate to change familiar, longstanding workflows. Start small with pilot projects that demonstrate how GIS simplifies tasks and improves outcomes. *Showing* success — not just talking about it — builds trust and momentum across the organization.

>> **Adapt workflows.** GIS often changes who does what and how tasks are completed. Provide clear guidance on new workflows and ensure that teams have the training needed to adapt. Tasks that once took weeks may now take only hours using GIS, but staff members need new skills to make that happen. Training not only empowers people to succeed but also inspires innovative solutions that benefit the whole team — a win for everyone!

>> **Ensure data security.** Work closely with IT staff to address data security concerns, especially when using cloud-based GIS solutions. Establish clear protocols to safeguard sensitive information and alignment with your organization's IT standards. When your IT department sees that your security goals align with theirs, they become better partners.

>> **Address resource allocation:** Implementing GIS isn't just about software. It requires investment in personnel, training, and infrastructure. Set a budget for these needs up front to avoid funding gaps and shortages.

TIP

As GIS becomes more integrated into your organization, keep your org chart and workflows updated regularly. This practice will help you track the evolving needs of your team and ensure that no one gets left behind.

Managing people problems

You may think that GIS is all about the technology, but it's nothing without the people involved in it. "Teamwork makes the dream work," a phrase made famous by John C. Maxwell in his 2002 book of the same name, may be overused but rings true. At the heart of every great GIS deployment is a cohesive, well-trained team. On the flip side, unless you've been stranded alone on a desert island your entire life, you know that people often pose some of the biggest challenges to organizational change. Here are common people-related challenges you may face and how to tackle them:

- » **Lack of support from leadership:** Every great team needs a champion. If organizational leaders lose interest in GIS, it may not reach its full potential. Keep leaders engaged by regularly showcasing GIS's value through success stories showing its positive impact on organizational goals.

- » **Unclear goals:** GIS initiatives without clear or aligned goals are destined to struggle. Instead of building your GIS around data and tools, focus on what you can do with GIS to support the organization and its clients. Make sure that all teams working with GIS understand how it supports broader organizational objectives.

- » **Insufficient training and support:** Lack of training programs and ongoing support can hamper GIS adoption and usage. Although many staff may already have GIS skills, they may need additional training tailored to organization-specific workflows. Document routine processes and ensure the availability of adequate training and resources for staff development.

- » **Lack of cooperation:** Turf wars, particularly in organizations with multiple GIS-related operations, can lead to dysfunction. Such conflicts often stem from a lack of communication. Work across teams to establish data standards, shared goals, and mutual incentives. Consider forming an organization-wide GIS user group to foster collaboration, discuss challenges, and work together on solutions.

- » **Resistance to change:** Many people simply don't like change. It can be scary, disrupting routines and requiring employees to learn new technologies. Address this trepidation head on by engaging staff early, involving them in decision-making, and providing plenty of training. Show them how GIS can make their work more efficient and impactful. When employees feel included, they're more likely to embrace change.

- » **Over-reliance on experts:** Some organizations lean too heavily on a small group of GIS experts, creating bottlenecks and limiting overall productivity. Cross-train staff to share knowledge and responsibilities, thereby building a more resilient and capable team.

REMEMBER

People problems aren't unique to GIS programs. Building a well-trained, collaborative team does more than just overcome challenges; it sets up your GIS program and organization for long-term success.

Planning for integration

Ken Eason, a pioneer in understanding how people and technology work together, created strategies for bringing new technology into organizations. Sometimes referred to as *Eason's principles*, these ideas involve making sure that the technology fits with both an organization's goals and the people who use it. Even though he first introduced the following principles back in 1988, they're still highly relevant in helping you introduce any technology into the workplace:

- » **Serve the organization's needs.** Make GIS a core part of your organization's strategy. Avoid relegating it to secondary tasks, like troubleshooting or specific mapping projects. Instead, position GIS as a critical tool for achieving goals and driving innovation.

- » **Empower employees to succeed.** Equip your team with the knowledge and skills they need to make GIS work. Provide training tailored to their work and encourage ongoing learning to keep up with new technologies.

- » **Plan for change.** Every organization has its own unique structure that evolves over time. Design workflows and systems that are adaptable to changing conditions – whether you're scaling operations, integrating new data sources, or adopting emerging technologies.

- » **Turn employees into stakeholders.** Make GIS a shared effort. Tie individual goals and team responsibilities to its success and involve employees in the decision-making process. When employees see GIS as a benefit rather than a burden, they're more likely to champion its use.

- » **Focus on goals and problem-solving.** Use GIS to address specific organizational needs, not just because it's the latest tech trend. Identify clear objectives, such as improving efficiency, enhancing decision-making, or driving customer satisfaction.

- » **Provide a system that fosters collaboration and buy-in.** Established, loyal employees are valuable assets, so give them every opportunity to integrate GIS into their work without feeling left out by new technologies.

- » **Prioritize employee well-being.** A system that makes workloads easier and reduces stress is far more likely to succeed. Design tools and processes that help employees work smarter, not harder. They'll enjoy their work more and likely perform better as a result.

- » **Invest in education.** Don't limit training to GIS staff. Managers, supervisors, and other decision makers should also understand GIS's role in the organization. GIS is continually advancing. The more knowledgeable your team is, the more effectively you can leverage GIS.

- » **Plan for growth.** GIS is constantly changing. Anticipate future needs, from software updates to expanded operations, and incorporate these into your long-term goals. An evolutionary approach minimizes workflow disruptions.

- » **Align with existing processes.** Change is easier when it feels familiar. The more closely new workflows align with the old, the quicker both employees and managers will buy in, which then leads to a smooth transition.

TIP Revisit your GIS strategy regularly to ensure that it stays aligned with organizational goals. Technology, needs, and priorities evolve, and your GIS should, too.

Building a Career in GIS

A career in GIS offers endless possibilities to explore, analyze, and solve spatial challenges across industries. You might find yourself managing parcel databases for a county government, creating interactive maps to assist with wildfire response, or using data analytics to help a major retailer reduce shipping costs. Whether you're just starting out or looking to advance, success comes down to understanding what it takes, committing to continuous learning, and building strong professional relationships.

Understanding what it takes to succeed in GIS

To succeed in the GIS industry, you need a solid foundation of education and skills. Most GIS professionals start with a bachelor's degree in geography, GIS, environmental science, urban and regional planning, or something related, but your specific focus depends on what you're aiming for. For example, a degree in business with a concentration in logistics can also open doors for a career in GIS. Regardless of your major, incorporating GIS courses in your studies is a must. Beyond education, a combination of practical skills and abilities will help you stand out when applying for that dream job.

Focus on these skill categories:

» **Technical expertise:** Proficiency with GIS software like ArcGIS Pro, QGIS, and web mapping platforms is essential. Experience with scripting languages like Python or R is a big plus. If coding isn't your thing, make sure that you understand the basics to help with automating tasks, building data queries, and enhancing analysis. (See Chapter 18 for a brief overview of common GIS programming languages and essential coding skills for all experience levels.)

» **Geospatial fundamentals:** You need a strong grasp of spatial data types, projections, and coordinate systems (refer to Chapter 2 for details on each of those topics). Hands-on experience with GPS data collection and processing is also valuable (both of which I cover in Chapter 7).

» **Data literacy:** As GIS increasingly intersects with big data and machine learning, skills like data wrangling (collecting and cleaning raw data), database

CHAPTER 19 **GIS in Organizations and Career Development** 311

management, and visualization techniques are more important than ever. (Find tips for working with data in Chapter 9.)

» **A problem-solving mindset:** GIS work requires critical thinking skills to solve spatial problems, identify spatial patterns, and interpret results. Successful GIS professionals approach challenges creatively, offering innovative, spatially driven solutions. (See Chapter 3 for more on the analytical skills and methods you use to interpret maps and spatial patterns in GIS.)

» **Communication skills:** Whether you're presenting complex GIS analysis to nontechnical audiences or collaborating with a team, the use of clear communication — both written and verbal — is a must. Likewise if your goal is to move into a leadership position.

» **Soft skills:** Alongside problem-solving and communication, skills like project management, attention to detail, and collaborative work with a team are vital for success as a GIS professional.

TIP

Create a GIS portfolio to build and showcase your skills. Include maps, projects, and analyses you've completed to highlight your technical expertise, problem-solving abilities, and creativity. Platforms like LinkedIn, StoryMaps, or a personal website are a great way to share your work with potential employers and collaborators. And don't forget to add links to interactive web maps or dashboards you've built!

Learning and growing in your GIS career

GIS is constantly evolving, and you should, too. Stay curious and never stop learning! Here are some strategies to keep growing with GIS:

» **Experiment with new tools and technologies.** Hands-on practice is the best teacher. Here are some suggestions for trying new tools and workflows:

- **Open the toolbox in your favorite GIS software,** look for a tool you've never used before, grab some sample data (see Chapter 21 for ideas), and start exploring and learning.

- **Volunteer to help with citizen-science projects** to work and learn alongside seasoned GIS professionals dedicated to giving back to the community. Check out Citizen Science GIS (www.citizensciencegis.org) or GIS Corps (www.giscorps.org) to get started.

- **Participate in the #30DayMapChallenge** (www.30daymapchallenge.com) or something similar to get your creativity flowing and build your cartographic and data visualization skills.

- **Take online courses.** Start with some of the many free online resources, such as Esri MOOCs (Massive Open Online Course), the QGIS Training Manual, and Coursera. Many organizations and universities provide free access to additional learning opportunities like Esri Academy or LinkedIn Learning. Also, be sure to check with your employer or university for additional learning opportunities.
- **Follow industry trends.** Stay up to date by attending conferences and webinars, subscribing to geospatial-related newsletters, and listening to podcasts such as The MapScaping Podcast, GISChat, and Geographers without Borders.

Growth doesn't always mean formal training. Every project, new tool, or challenge is an opportunity to expand your GIS expertise.

Networking and professional growth

In GIS, connections matter. Networking opens doors to new opportunities, keeps you in the loop on industry trends, and surrounds you with a community of passionate professionals who "get it." And don't underestimate certifications; whether they're professional or technical, certifications can give you a real edge in career advancement. Here are some ways to grow your network and elevate your GIS game:

- **Join professional organizations.** Connect with groups like The Geospatial Professional Network (GPN, formerly URISA), ASPRS (American Society for Photogrammetry and Remote Sensing), and NaCIS (North American Cartographic Information Society) to access conferences, educational resources, and networking communities. These organizations and others like them are gateways to people and tools that can shape your career.

A lot of great geospatial organizations exist, and many are targeted to specific audiences. Some of these specialized organizations include Women in GIS, The Society of Conservation GIS, and NorthStar of GIS. But don't stop there: Search online or ask a colleague, mentor, professor, or fellow GIS professional for recommendations to find an organization that resonates with your interests and goals.

- **Earn a professional certification.** Offered by the GIS Certification Institute (GISCI), the GIS Professional (GISP) certification is the most widely recognized professional certification in GIS. Earning your GISP not only proves you have the experience and smarts to be a GIS professional but also serves as a badge of honor that shows your commitment to the profession. (See the nearby sidebar "Why bother with a professional certification?" for more about the benefits of certification.)

CHAPTER 19 **GIS in Organizations and Career Development** 313

- » **Get a technical certification.** Technical certifications are how you show off your mastery of specific tools. The ArcGIS certifications from Esri and the ASPRS credentials in Photogrammetry, Remote Sensing, Lidar, and Unmanned Aerial Systems (UAS) are just a few options that can boost your expertise in a specialized area.

- » **Attend conferences and events.** Nothing beats connecting with people in real life. Industry events like the Esri International User Conference, OpenStreetMap's State of the Map, GPN's GIS-Pro, or local GIS meetups are where great things happen. They're perfect for learning, showcasing your work, and rubbing elbows with industry leaders. And who knows — you might just find your next mentor or collaborator!

If you're an introvert, don't worry — you don't have to work the room like a glad-handing politician. Start small by attending a local event that interests you, connecting with other attendees beforehand in an online community, or setting a goal to have just one meaningful conversation during the event. Many conferences also offer quiet spaces where you can recharge between sessions, or you can slip away to find your own quiet spot whenever needed.

- » **Participate in online communities.** Join the conversation with other GIS folks on LinkedIn, Bluesky, or other platforms. Look for hashtags like #GISChat to share ideas, ask questions, or enjoy some great GIS memes.

- » **Give back.** Volunteering is a win-win. Get involved with organizations like GISCorps to sharpen your skills, contribute to meaningful projects, and connect with others who are just as passionate about GIS as you are. Bonus: Volunteering looks great on your resume and might lead to your next big opportunity.

Networking and professional growth in GIS go hand in hand. So take that first step and join a group, attend an event, or lend your expertise to a good cause. You never know where it might lead!

WHY BOTHER WITH A PROFESSIONAL CERTIFICATION?

Professional certifications show employers and peers that you're serious about your career and have the expertise to prove it, and the GISP certification can be your next step in showcasing your expertise and dedication to GIS. But you may be wondering whether the GIS Professional (GISP) certification is right for you.

If you're well-established and happy in your current position, obtaining this certification may not make a difference. But if you're early in your career or looking to stand out in the growing GIS field, GISP certification can give you an edge. With more employers listing it as a preference or requirement for mid- to senior-level roles, having "GISP" after your name can boost your credibility.

Here are the three main requirements for a GISP certification:

- **Have an approved portfolio.** The GISP portfolio isn't a compilation of your maps and apps; it's more like a resume. A completed portfolio shows that you meet the requirements for
    - **Education:** A degree or equivalent geospatial-related coursework
    - **Experience:** At least four years of GIS work
    - **Contributions to the profession:** Activities like volunteering, presenting at conferences, or professional organization involvement
- **Pass the exam.** The GISCI Geospatial Core Technical Knowledge Exam tests foundational knowledge of GIS concepts, tools, and best practices. It's software agnostic, so you aren't tested on how to use any specific GIS software.
- **Abide by the Code of ethics.** GISPs agree to uphold ethical standards to ensure that their work benefits society and reflects the profession's integrity.

The certification process is flexible. Start tracking your experience and contributions early by creating an account on the GISCI website (www.gisci.org). You can take the exam whenever you're ready, and you don't have to complete the steps in any particular order. To get started, visit the GISCI website for resources or guidance.

Exploring career paths in GIS

The GIS field is as diverse as the data it analyzes. From managing city infrastructure to tracking wildlife migration to identifying prime locations for business expansion, GIS professionals work in almost every industry imaginable. Understanding the different career paths can help you focus your education and find a role that fits your skills and interests. Here are some major focus areas to consider:

>> **Government (the public sector):** GIS is a cornerstone of government operations. City planners use GIS to map land use and design zoning policies, emergency responders rely on it for disaster preparedness and response, and transportation departments manage roadway infrastructure while maintaining network datasets. Whether you work for local, state, or federal government, GIS offers countless opportunities to make a difference.

» **Environmental science and conservation:** If you love the great outdoors, you can put your GIS skills to work in this field in a wide range of ways. GIS plays a huge role in monitoring ecosystems, managing natural resources, and conserving biodiversity. You might track endangered species, assess climate change effects, or plan sustainable development projects. Roles in environmental consulting firms, nonprofits, or government agencies let you put GIS to work for Mother Earth.

» **Urban planning and infrastructure:** GIS helps urban planners visualize data to shape how cities grow and function. From designing public transit systems to planning utilities, GIS professionals in this field solve spatial problems that improve quality of life for residents.

» **Business and logistics:** Companies increasingly rely on GIS to gain a competitive edge. Retailers analyze customer demographics to choose new store locations; delivery services optimize routes to save time and fuel; and marketing teams use spatial analysis to target campaigns. If you're interested in business applications, a GIS career in the private sector may be for you.

» **Technology and software development:** If you have a knack for coding, consider a career in GIS technology. Developers create mapping applications, design spatial databases, and innovate geospatial tools that expand what GIS can do. Although GIS software companies like Esri, Mapbox, or geospatial startups are common destinations, developers are needed across all industries, not just technology firms.

» **Academia and research:** If your interests incline toward teaching or advancing GIS knowledge, you may find a home in a career in academia, focusing on education, research, and innovation. From developing new spatial analysis techniques to mentoring the next generation of GIS professionals, this path is perfect for those who love to learn and share knowledge.

REMEMBER

This section's list of career paths is by no means exhaustive. You can find GIS just about anywhere! The key is to find an area that excites you, and then you can develop the skills and experience to succeed in that niche.

The Part of Tens

IN THIS PART . . .

Explore ten GIS software options to find a GIS solution that fits your needs and budget.

Discover ten GIS data sources to find data to use in your next GIS project.

Take a peek into what's next for GIS with a list of ten trends to watch.

IN THIS CHAPTER

» Discovering some of the most widely used GIS platforms

» Exploring unique features of different GIS software

» Choosing the right GIS software for your needs

Chapter **20**

Ten GIS Software Options

GIS software has come a long way since its beginning in the 1960s, when Dr. Roger Tomlinson introduced the Canada Geographic Information System to support land-use management and resource monitoring. Today, GIS is everywhere — from mapping endangered species habitats to optimizing delivery routes and even predicting impacts of climate change.

Whether you're a professional in urban planning, an environmental scientist, or a curious beginner, the right GIS software can make all the difference. This chapter introduces you to ten of the most widely used GIS platforms, each with its own strengths. Whether you need a powerhouse like ArcGIS Pro, an open-source solution like QGIS, or a specialized tool like Google Earth Engine, there's something here for every use and budget.

Choosing the right GIS software can feel overwhelming, but it boils down to your goals, technical expertise, and budget. No matter which GIS software you choose, be sure to explore its tutorials and community resources. GIS is as much about learning the tools as it is about solving problems with them!

ArcGIS Pro

https://www.esri.com/en-us/arcgis/products/arcgis-pro/overview

ArcGIS Pro is a comprehensive GIS platform developed by Esri, offering tools for mapping, spatial analysis, and 3D modeling. It integrates with ArcGIS Online for sharing and collaboration and supports extensions for specialized tasks.

» **Key features:**

- Provides a comprehensive GIS toolset with advanced spatial analysis, cartography, and 3D modeling
- Lets you share and collaborate effortlessly with ArcGIS Online
- Includes support for Python scripting through ArcPy and specialized statistical analysis through the R-ArcGIS Bridge
- Offers an extensive library of extensions for advanced tasks like network analysis, business analysis, and image classification

» **Best for:** ArcGIS Pro is a good choice if you need a full-featured GIS solution that supports everything from map-making to complex geospatial analytics. It's particularly useful in enterprise environments where advanced spatial analysis and data integration are required.

» **Limitations:**

- **Cost:** ArcGIS Pro operates on a subscription-based licensing model, which may be expensive for small organizations or individual users.
- **Learning curve:** ArcGIS Pro's extensive features can make the interface difficult to navigate and complex tools difficult to understand, requiring significant time and training to master.

» **About Esri:** Founded in 1969 as Environmental Systems Research Institute, Esri is the global leader in GIS software. Esri offers a vast suite of products and a strong commitment to education and community-building.

QGIS

https://www.qgis.org/

QGIS (formerly Quantum GIS) is an open-source GIS application that provides tools for creating maps, analyzing spatial data, and managing geospatial databases. It supports integration with other open-source tools such as GRASS GIS, GeoServer, and PostGIS (the spatial database extension for PostGreSQL).

» **Key features:**

- Includes a wide array of intuitive geospatial tools for creating maps, editing layers, and performing spatial analysis
- Is highly customizable with plug-ins for specific workflows like hydrological modeling, transportation analysis, and more
- Supports a large range of vector, raster, and database formats, including shapefiles, geodatabases, GeoJSON, and GeoPackage
- Runs on Linux, Unix, macOS, and Windows

» **Best for:** QGIS is a strong option if you're a student, researcher, or part of a small organization looking for a cost-effective GIS tool. It's also ideal for anyone needing an open-source solution that integrates well with other tools.

» **Limitations:**

- **Support:** As an open-source project, QGIS relies on community support, which may not meet the needs of all users.
- **Performance:** QGIS may experience slower performance when handling very large datasets or complex spatial analyses compared to some commercial GIS software that is optimized for high-performance computing.

» **About QGIS:** QGIS is an open-source project developed by a dedicated global community of developers, educators, and users. Its open-source nature and active community means that it's constantly evolving, giving you access to the latest tools and resources at no cost.

MapInfo Pro

https://www.precisely.com/product/precisely-mapinfo/mapinfo-pro

MapInfo Pro, by Precisely, is a GIS application designed for mapping and spatial analysis, commonly used in business and marketing.

» **Key features:**

- Uses an intuitive interface designed for quick analysis and visualization
- Provides strong support for demographic and business data integration

- Includes tools for thematic mapping, spatial analysis, and geocoding
- Offers seamless data sharing through various formats and platforms

» **Best for:** MapInfo Pro is designed for business analytics and marketing professionals who need spatial insights but don't have a strong GIS background.

» **Limitations:**
  - **Cost:** MapInfo Pro requires a paid license, which may be limiting if you're on a tight budget.
  - **Fewer advanced tools:** MapInfo Pro lacks some advanced spatial modeling and geostatistical tools found in software like ArcGIS Pro or GRASS GIS.

» **About Precisely:** Precisely, a company specializing in data integration, acquired Pitney Bowes' MapInfo software and data business in 2019. Although its primary focus is on data management and analytics, Precisely also operates in the geospatial technology sector through its MapInfo products and related offerings.

GRASS GIS

https://grass.osgeo.org/

GRASS GIS (Geographic Resources Analysis Support System) is a popular open-source software for advanced spatial modeling and environmental analysis. If you're a researcher or an environmental scientist, GRASS offers tools for geospatial data management and analysis, image processing, and spatial modeling.

» **Key features:**
  - Extensive capabilities for data analysis, optimized for large-scale analysis on various hardware configurations
  - Advanced modeling for terrain, hydrology, and ecosystems
  - Built-in temporal framework and Python API, enabling advanced time series analysis and rapid geospatial programming
  - Seamless integration with other open-source tools like QGIS

- **Best for:** GRASS GIS is focused on terrain and ecosystem modeling, hydrology, imagery processing, and large-scale data projects, particularly in research and environmental science.
- **Limitations:**
  - **User interface:** The GRASS GIS interface is less intuitive compared to other GIS software, creating a steeper learning curve.
  - **Data imports:** It requires data to be imported into the GRASS GIS native file format, which can add additional processing time when working with external datasets.
- **About GRASS GIS:** GRASS GIS was developed by the U.S. Army Corps of Engineers in 1982. Today, it's maintained and managed by a global steering committee and is backed by the Open-Source Geospatial Foundation (OSGeo).

AutoCAD Map 3D Toolset

```
https://www.autodesk.com/products/autocad/included-toolsets/autocad-map-3d
```

AutoCAD's Map 3D toolset provides access to CAD and GIS data, giving you the ability to directly use, edit, and manage geospatial data within a CAD environment. This toolset bridges the gap between engineering and GIS.

- **Key features:**
  - Combines GIS data with CAD workflows for precise design
  - Provides tools for creating, editing, and managing spatial data with industry-specific templates
  - Gives you direct access to spatial data, including streamlined data flow between ArcGIS and Map 3D for editing and mapping
  - Includes strong support for geospatial data visualization and thematic mapping.
- **Best for:** Map 3D is great for engineers, architects, and planners who need to combine GIS functionality with detailed CAD designs for infrastructure projects and spatial analysis.

- **Limitations:**
  - **Cost:** AutoCAD Map 3D is available only as part of an AutoCAD subscription, which may not be cost-effective if you need only GIS capabilities.
  - **Complexity:** If you're familiar with CAD, adapting to the GIS functionalities may take time (and vice versa).
- **About Autodesk:** Known for AutoCAD, Autodesk provides software solutions for architecture, engineering, and construction. Autodesk offers a diverse product suite that also includes tools for product design, manufacturing, media, and entertainment.

Google Earth Engine

https://earthengine.google.com/

Google Earth Engine is a cloud-based platform for processing and analyzing large geospatial datasets, primarily satellite imagery and environmental data.

- **Key features:**
  - Access to a vast cloud-based repository of geospatial datasets, including satellite imagery and climate data
  - Cloud-based processing for handling large-scale datasets quickly
  - JavaScript and Python APIs for advanced analysis and automation
  - Powerful visualization tools for interactive mapping and time-series analysis
- **Best for:** Google Earth Engine helps you with environmental modeling, disaster management, and large-scale geospatial analysis, particularly if you often deal with massive datasets or remote-sensing workflows.
- **Limitations:**
  - **Internet dependency:** Because it's cloud-based, Google Earth Engine requires a stable internet connection for accessing and analyzing data.
  - **Data privacy:** When handling sensitive information, you need to consider data privacy and compliance when using this cloud-based application.
- **About Google:** Google is not just a verb; it's a large technology company that, while known for its popular search engine, also provides software solutions for data processing, analytics, and cloud computing.

GeoServer

https://geoserver.org/

GeoServer is an open-source server application designed for publishing maps and sharing geospatial data using web services that are compliant with the Open Geospatial Consortium (OGC) standards.

» **Key features:**
- Supports a wide range of geospatial formats and standards, including Web Mapping Service (WMS), Web Feature Service (WFS), and Web Coverage Service (WCS)
- Provides flexible styling options for maps and data layers using Styled Layer Descriptors (SLD)
- Integrates with other open-source tools like QGIS and PostgresSQL/PostGIS
- Is scalable for small projects or enterprise-level deployments

» **Best for:** GeoServer is helpful for organizations or teams looking to share geospatial data online or integrate with open-source ecosystems.

» **Limitations:**
- **Technical expertise:** GeoServer requires knowledge of web services and server configurations, which may be challenging if you're unfamiliar with managing GIS servers.
- **Performance:** Handling high-demand services or large datasets may require the use of optimization tools or additional resources.

» **About GeoServer:** Built by contributors from a global community of developers and users, GeoServer is free, open-source software that uses the open standards set forth by the Open Geospatial Consortium (OGC).

ERDAS IMAGINE

https://hexagon.com/products/erdas-imagine

ERDAS IMAGINE is a software tool for processing and analyzing raster data.

» **Key features:**
- Advanced tools for image classification, enhancement, orthorectification, reprojection, and interpretation

- Provides tools for raster analysis, point cloud processing, and integration with vector data
- 3D visualization and spatial modeling capabilities
- Scalable options for enterprise-level image processing

» **Best for:** ERDAS IMAGINE is commonly used in remote sensing, defense, and environmental modeling for processing satellite and aerial imagery.

» **Limitations:**

- **Cost:** ERDAS IMAGINE is a premium software with a pricing model that may be out of reach for many budgets.
- **Learning curve:** The interface and workflow can be complex if you're unfamiliar with remote sensing software.

» **About Hexagon:** Hexagon is a software technology company specializing in digital reality solutions. They offer a variety of products and services to support sustainable, autonomous solutions.

Global Mapper

https://www.bluemarblegeo.com/

Global Mapper offers GIS tools with a user-friendly interface. It's known for supporting terrain analysis, LiDAR processing, and a variety of geospatial data formats.

» **Key features:**

- Supports more than 300 file formats for seamless data interoperability
- Includes tools for terrain analysis, 3D visualization, and vector data management
- Provides an easy-to-use interface for quick data processing and analysis
- Optional LiDAR module for advanced point cloud processing

» **Best for:** Global Mapper is a great tool for small businesses, field users, or anyone needing affordable, versatile tools for mapping, terrain analysis, and LiDAR processing without having to deal with a steep learning curve.

» **Limitations:**

- **Limited toolset:** Global Mapper is strong for LiDAR and terrain analysis but lacks some spatial modeling and statistical tools, such as advanced visibility analysis (viewshed and line-of-sight calculations).

- **User base:** Compared to ArcGIS Pro and QGIS, Global Mapper has a smaller user base in research, government, and enterprise GIS environments.

» **About Blue Marble Geographics:** Blue Marble Geographics is a GIS software company founded in 1993. In addition to the Global Mapper GIS products, they are known for their coordinate conversion software and LiDAR data processing tools.

OpenCities Map

https://www.bentley.com/software/opencities-map/

OpenCities Map provides GIS tools for infrastructure planning, design, and management. It is commonly used by utility companies, urban planners, and infrastructure managers.

» **Key features:**
- Advanced geospatial analysis and map creation tailored to infrastructure needs
- Integration with CAD, BIM (Building Information Modeling), and GIS datasets
- Tools for managing utilities, land parcels, and city planning projects
- Support for spatial querying, topology, and thematic mapping

» **Best for:** Infrastructure managers, utility companies, and urban planners will appreciate the GIS tools in OpenCities Map that integrate seamlessly with engineering and construction workflows.

» **Limitations:**
- **Narrow focus:** OpenCities Map is specialized for infrastructure and utilities management, making it less suitable for general-purpose GIS tasks like environmental analysis or land-use planning.
- **Learning curve:** OpenCities Map integrates with CAD and BIM, which can be a limitation if you're unfamiliar with engineering workflows.

» **About Bentley Systems:** Bentley Systems provides solutions for designing, constructing, and operating infrastructure assets. Their products emphasize integration between GIS and engineering disciplines.

IN THIS CHAPTER

» **Exploring reliable GIS data sources**

» **Understanding global and regional GIS data options**

» **Finding data for specific projects and analysis**

Chapter **21**

Ten GIS Data Sources

GIS is nothing without data. Whether you're creating detailed maps, performing advanced analysis, or visualizing global trends, reliable data is the foundation of every GIS project. For some projects, you may find yourself relying on data that either you or a colleague collected in the field (as I describe in Chapter 7), but more often than not, most of the data that you need for your geospatial projects already exists, which can save you time and effort if you locate it.

The ten GIS data sources included in this chapter provide access to high-quality geospatial data, including global, national, and thematic datasets available for your use in a variety of applications. This list is just the tip of the iceberg; many, many more data sources are out there! In fact, I had trouble narrowing my list down to just ten, but I've selected a variety of useful sources to get you started on a data discovery journey.

TIP

To help you evaluate GIS data sources, ask these questions:

» **Does this dataset include metadata?** Metadata explains where the dataset came from, when it was collected, and how it's structured, which is critical information for ensuring compatibility and quality.

» **Do the data meet your project's needs?** Think about resolution, accuracy, and whether the format works with your GIS software.

- » **Are the data reliable?** Look for authoritative or well-known sources to ensure quality, accuracy, and reliability.

- » **Does the dataset have licensing restrictions?** Some data are free to use, but others may require permissions or fees.

- » **Does the source provide support or tutorials?** Resources like guides, technical documentation, FAQs, or community forums can be invaluable if you're just getting started with a new and unfamiliar dataset.

Metadata is your best friend when it comes to evaluating GIS data. Always check for metadata to understand the origins, limitations, and structure of the data. It's key to building a reliable, useful, and informative GIS project.

Many modern GIS data sources now offer an application programming interface (API), making it easier than ever to automate data updates, integrate datasets into your custom workflows, and analyze large datasets programmatically. If you're comfortable with a bit of coding, APIs can be a game-changer! Check out Chapter 18 for an introduction to APIs. The data provider's documentation may also include tutorials and examples for using APIs. In this chapter, I note those that provide API access (as of this writing, anyway).

OpenStreetMap

OpenStreetMap (OSM) is the Wikipedia of mapping. It's a global, community-driven platform where anyone can contribute and edit geospatial data. With millions of contributors worldwide, OSM provides free and detailed vector data that you can adapt for countless GIS projects.

- » **Best for:** General-purpose mapping, urban planning, transportation analysis, and projects requiring regularly updated geographic data.

- » **How to access:** Go to https://OpenStreetMap.org. You can download data from the site using tools like Geofabrik, the Humanitarian OpenStreetMap Team (HOT) export tool, or the Overpass API.

  The Overpass API is perfect for customizing your data downloads, whether you need detailed road networks, building footprints, or specific points of interest. Bonus: It has a tutorial to get you started!

- **Limitations:** The quality of the data can vary depending on the region. Urban areas are often very detailed, but coverage in rural or remote areas may be patchy.

ArcGIS Online and Living Atlas

Esri's ArcGIS Online and Living Atlas of the World data portals offer a treasure trove of authoritative geospatial data, including high-resolution imagery, demographic layers, and environmental datasets. If you're using Esri's tools like ArcGIS Pro, these resources are incredibly helpful for building maps and conducting analyses.

- **Best for:** Professionals using Esri products who need pre-packaged, authoritative datasets for analysis, mapping, and sharing.

- **How to access:** Log in to https://arcgis.com with an Esri account to begin exploring the data. Go to https://livingatlas.arcgis.com/ to browse data available in ArcGIS Living Atlas of the World.

  The REST API at ArcGIS Online makes it easy to fetch, analyze, and even update data directly in your applications. Check Esri's developer website for plenty of examples and guides.

- **Limitations:** Although you can create a free public account on ArcGIS.com (for noncommercial uses), access to some datasets requires an ArcGIS subscription. Integration with non-Esri platforms can be tricky.

Natural Earth

Natural Earth is a free, public-domain resource that's all about simplicity and quality. Offering global-scale vector and raster datasets, it's a favorite among cartographers for making beautiful and informative maps.

- **Best for:** General-purpose mapping, cartography, and educational projects. Perfect for lightweight GIS tasks and quick map creation.

- **How to access:** Visit https://naturalearthdata.com to download the data directly.

» **Limitations:** The datasets are generalized for small-scale maps (large geographic areas), so they're not suitable for local or highly detailed analyses.

IPUMS NHGIS

IPUMS National Historical GIS (NHGIS) is a popular resource for harmonized U.S. Census data and boundary files. Whether you're analyzing population trends or studying housing patterns, NHGIS integrates demographic data with GIS-ready formats for easy use.

» **Best for:** Population studies, housing analyses, and projects requiring historical and harmonized demographic data.

» **How to access:** Create a free account at https://nhgis.org to browse and download the data. The IPUMS API provides another way for you to explore and retrieve NHGIS data extracts.

Although it's not specific to IPUMS NHGIS, you can also download U.S. Census data directly through the Census APIs using Python and R programming tools like the TidyCensus package (see https://walker-data.com/tidycensus/).

» **Limitations:** Data are provided in shapefile format, and you must manually join datasets using a GISJOIN field for analyses.

Copernicus Data Space Ecosystem

Copernicus, The European Union's Earth observation program, delivers free, high-resolution satellite imagery and Earth observation data through tools like the Copernicus Browser and APIs. Copernicus is an essential resource for remote-sensing enthusiasts.

» **Best for:** Environmental monitoring, land cover analysis, and disaster response.

» **How to access:** Visit https://dataspace.copernicus.eu/ to access the data. In addition to its user-friendly web interface, Copernicus provides an API for accessing satellite imagery and Earth observation data programmatically. Check the API documentation on its website for more details.

» **Limitations:** Processing raw imagery requires remote sensing expertise, making it less beginner friendly.

USGS EarthExplorer

If you need remote sensing or environmental data, EarthExplorer is a must. Managed by the U.S. Geological Survey (USGS), this platform provides free access to extensive remote sensing data, including Landsat imagery, aerial photography, and elevation models.

» **Best for:** Remote sensing projects, environmental analysis, and historical imagery studies.

» **How to access:** Register and explore at https://earthexplorer.usgs.gov/. The USGS EarthExplorer API and LandsatLook Viewer API allow for efficient bulk downloads or repetitive queries, saving you precious time when working on large-scale remote sensing projects.

» **Limitations:** Registration is required to download data, and some data may have access restrictions or require expertise for use.

SEDAC

NASA's Socioeconomic Data and Applications Center (SEDAC) bridges the gap between Earth science data and social science, offering datasets that integrate population, climate, and sustainability factors.

» **Best for:** Sustainable development, urban planning, and climate resilience projects.

» **How to access:** Visit https://www.earthdata.nasa.gov/centers/sedac-daac. SEDAC also provides some datasets through NASA's Earthdata APIs, allowing for streamlined access to socioeconomic and environmental data.

» **Limitations:** You need to create a free Earthdata account to access and download data. The website and tools can be complex for new users to navigate.

WorldClim

WorldClim offers free global climate and weather data, perfect for ecological modeling and agricultural planning. With high-resolution climate variables, it's a valuable resource for understanding past, present, and future climate trends.

- » **Best for:** Ecological modeling, agricultural planning, and climate impact studies.
- » **How to access:** Visit https://worldclim.org to download datasets.
- » **Limitations:** Data resolution varies by dataset, and global coverage may miss local details.

GeoPlatform.gov

The U.S. federal government's GeoPlatform is a collaboration hub for authoritative geospatial data, offering tools for visualizing and analyzing topics like infrastructure, public health, and the environment

- » **Best for:** U.S.-based projects requiring federal data or collaboration with government agencies.
- » **How to access:** Visit https://geoplatform.gov to explore and download data. GeoPlatform also supports programmatic access through APIs for many of its datasets.
- » **Limitations:** Most datasets are U.S.-centric, limiting global applications.

INEGI

Mexico's INEGI, the National Institute of Statistics, Geography, and Data Processing, is a treasure trove of geospatial data, providing detailed cartography, demographics, and environmental datasets for the country.

- » **Best for:** Projects focusing on Mexico's demographics, land use, or environmental studies.

» **How to access:** Visit https://inegi.org.mx for datasets and tools. INEGI's Datos Abiertos API (Open Data API) lets you query datasets like population density maps or detailed land use classifications, which are ideal for researchers who need highly localized data.

» **Limitations:** Resources are primarily in Spanish, which may require translation for international users.

WARNING

DON'T PAY FOR WHAT YOU CAN GET FOR FREE!

Not all GIS data are created equal, and sometimes companies sell "value-added" data, or datasets they claim to have enhanced. Although these datasets can be useful, it's worth pausing to ask yourself, "Is this data really better, or can I find something similar for free?"

Here's how to decide whether value-added data is worth the cost:

- **Check for complete metadata.** Metadata reveals where the data came from, how it was processed, and its limitations. If the company doesn't provide metadata, it's a red flag!

- **Do a quick search.** Many public datasets are free or low cost. Make sure that the vendor isn't charging for something you can download for free or less cost elsewhere.

- **Ask around.** GIS communities, university faculty, and user groups can often point you to reliable data sources or share insights about a vendor's reputation.

- **Think long term.** Quick fixes might seem appealing, but poor quality data can lead to headaches and extra costs down the road. Take your time to find good-quality data.

If you decide to pay for data, ask for a contract that clearly outlines how the data has been enhanced and ensure that it meets your project's needs.

> **IN THIS CHAPTER**
> » Exploring the impacts of GeoAI, 3D modeling, and cloud GIS
> » Understanding the importance of geospatial ethics and privacy
> » Discovering how citizen science and open-source tools are democratizing GIS

Chapter **22**

Ten GIS Trends to Watch

The world has changed a lot since the first edition of this book was published in 2009. Technology has leapt forward in ways that have reshaped industries and everyday life, from the rise of smartphones and 5G networks to the proliferation of Internet of Things (IoT) devices and breakthroughs in artificial intelligence (AI). These advancements haven't just transformed how we live and work; they've profoundly impacted GIS and how we interact with geospatial data.

Even if you're Nostradamus, predicting the future is a tricky business (and, honestly, a bit risky to do in a book that may be around for years to come). Some of the trends I highlight in this chapter may fade into obscurity; others may define the next era of GIS innovation. But based on where technology has been and where it seems to be heading, I've pulled together ten GIS trends worth keeping an eye on. In this chapter, you discover trends that can shape the tools you use, the data you analyze, and how you solve problems in the years to come.

Integration of AI and Machine Learning

AI and machine learning are transforming GIS by automating tasks like image classification, feature extraction, and real-time analysis. In fact, there's even a term for this automation: GeoAI. With GeoAI (short for Geospatial Artificial Intelligence), you can identify areas at risk for wildfires, determine routes based on real-time traffic data, and predict urban growth patterns more effectively and efficiently than ever.

GeoAI combines GIS and advanced computing methods like deep learning and predictive analytics to analyze and interpret spatial data. These techniques help uncover patterns and relationships that may not be obvious through traditional methods. By training models with satellite imagery, sensor data, and geospatial datasets, GeoAI can detect land cover changes, identify infrastructure damage after disasters, or even generate high-accuracy geospatial data layers from unstructured data sources.

GeoAI is not a futuristic concept; it's already mainstream. The maker of the ArcGIS suite, Esri, integrates GeoAI into image classification, object detection, and predictive modeling through this suite. QGIS supports GeoAI capabilities by way of plug-ins like the Semi-Automatic Classification Plugin (SCP) for land cover classification. Meanwhile, Google Earth Engine incorporates machine learning tools for environmental monitoring and pattern detection.

Although the advancement of GeoAI is a game-changer, here are some important factors to consider when using it:

» **Data quality:** GeoAI relies on high-quality data to work effectively. The old adage "garbage in, garbage out" applies here. Your model is only as good as the data you feed it.

» **Your understanding of the process:** Knowing how your GeoAI model arrived at the results is critical. You control the workflow and set parameters, so make sure they are well-defined and suited for your goals.

» **Sufficient computing power:** The nature of GeoAI work is to automate tasks that often require large datasets and complex algorithms. Ensure that your hardware and cloud resources are up to the task.

» **Ethical concerns:** Data privacy, algorithmic bias, and misuse of information are all risks with GeoAI. You can address these issues by checking your data for biases, developing transparent workflows, implementing robust data privacy practices, and engaging with stakeholders.

TIP

Start small. Experiment with tools like Google Earth Engine before scaling up to complex GeoAI workflows.

3D GIS and Augmented/Virtual Reality

GIS has moved beyond 2D maps into immersive 3D environments that model cities, landscapes, and building interiors. Augmented reality (AR) and virtual reality (VR) expand 3D GIS even further by providing tools for city planners to visualize skylines before construction and emergency responders to simulate disaster scenarios. Tools like Unity (popular for game development) and Esri's CityEngine enable detailed 3D models and interactive AR/VR experiences, often requiring collaboration with modeling, programming, or design experts.

If you're getting started with 3D GIS, here are a few key considerations:

- **Sufficient computing power:** 3D GIS is data intensive and demands significant computing power. Make sure that your hardware and software are equipped to handle large datasets and complex visualizations.

- **Data quality:** Accuracy is critical. Ensure that your 3D data aligns well with real-world features to provide a reliable and useful AR/VR experience.

- **User experience:** Designing 3D models isn't just about the data but also the data's usability. Think about how users will navigate and interact with your model. Choose an appropriate level of detail: Will a simple model suffice, or is a high-resolution model necessary? Pay attention to colors, textures, and annotations to ensure that your 3D environment effectively communicates your message.

TIP

If you're new to 3D, practice converting existing 2D data into simple 3D models before diving into AR/VR. For example, if you have a building footprints layer, try adding elevation values to represent building height; then use GIS tools to create basic 3D models. In ArcGIS Pro, the Extrusion tools let you lift flat features into 3D based on height attributes. In QGIS, the Qgis2threejs plug-in provides an easy way to generate simple 3D visualizations.

Cloud-Based GIS and SaaS Solutions

GIS is rapidly shifting to the cloud, where data, tools, and workflows are hosted online. With cloud-based platforms like Esri's ArcGIS Online and open-source options like GeoServer, GIS professionals can access their projects from anywhere with an internet connection. Software as a Service (SaaS) models are growing, providing scalability, ease of collaboration, and reduced hardware costs. For example, you can upload data, perform analysis, and share maps with colleagues or the public — all without downloading a thing.

Cloud GIS is revolutionizing the field, but here are some key points to consider:

- **Have a reliable internet connection:** Cloud GIS depends on internet access, so a fast, stable connection is a must. Offline options may be limited, so have a backup plan if you work in areas with spotty service.

- » **Account for costs and subscriptions:** Saas platforms typically operate on subscription models, which can be cost-effective for some but add up over time. Be sure you understand the pricing tiers and whether you'll need paid extras like premium data or advanced tools.

- » **Ensure your data's privacy and security:** Storing proprietary data or sensitive information — such as Personally Identifiable Information (PII) or Protected Health Information (PHI) — in the cloud raises legal and ethical concerns. Many countries, including the United States, have strict regulations that govern how this type of information must be handled. Always review the provider's security protocols, consider encryption or access controls, and ensure compliance with applicable laws before storing confidential data in cloud-based GIS platforms.

Many platforms offer free trials or basic plans, so test their features before committing to a paid subscription.

Expansion of Remote Sensing Capabilities

Advances in satellite and drone technologies are revolutionizing remote sensing, enabling you to capture higher-resolution imagery, revisit locations more frequently, and explore new sensing capabilities like hyperspectral imaging, which analyzes a wide range of wavelengths to reveal details invisible to the human eye. Remotely sensed data is essential for applications like monitoring deforestation, assessing natural disasters, and precision farming.

To get the most out of remote sensing, be aware of the following:

- » **You may need specialized tools and knowledge:** Remote sensing data often requires processing software like ENVI, Google Earth Engine, or ArcGIS Pro, as well as skills in working with raster data. If working with remote sensing interests you, learning the basics of raster data (see Chapter 8) and remote sensing workflows can get you started on the right track.

- » **You want to stay current with technology:** Advances in satellite constellations and the increasing availability of affordable drones have democratized access to remote sensing, so explore these options to match your project's needs and interests.

Platforms like Copernicus (see Chapter 21) and Sentinel Hub offer free imagery and tools, making it easier to experiment with remote sensing data on a budget.

Evolving Geospatial Data Privacy and Ethics

As geospatial data becomes more ubiquitous, so do concerns about privacy and ethical use. For instance, location data collected by mobile apps can reveal sensitive personal details, and unsecured datasets can expose users to fraud and surveillance. On a broader scale, these issues also extend to how governments and organizations use location for decision-making.

Here's what you should keep in mind:

- **Understand the laws:** Regulations like General Data Protection Regulation (GDPR) in Europe and California Consumer Privacy Act (CCPA) in the U.S. set rules for data protection and transparency. Be sure to familiarize yourself with the requirements in your area.

- **Build trust:** Be transparent about how you collect, store, and use location data. Create workflows that minimize bias, protect individual privacy, and promote data security.

- **Ethical implications:** Always ask yourself how your geospatial data can be misused. Can it harm marginalized communities or violate personal privacy? Awareness is the first step in preventing unintended consequences.

Ethical data handling is becoming more of a factor in funding and partnerships, so make it a priority in your work.

Growth of Open-Source GIS and Data

Open-source GIS platforms like QGIS and open-data initiatives like OpenStreetMap are leveling the playing field, making GIS accessible to students, small organizations, and anyone on a limited budget. Open-source GIS tools empower users with customizable features, and open data initiatives provide free access to geospatial datasets.

Here's what you need to know about these open-source trends:

- **Community innovation drives them.** Open-source GIS tools evolve rapidly, thanks to engaged volunteer communities of developers and users. Plug-ins, forums, and tutorials are just a click away.

CHAPTER 22 **Ten GIS Trends to Watch**

- » **Technical know-how is required.** Open-source tools are incredibly powerful, but they may require more self-directed learning compared to commercial software. Although structured training opportunities may be fewer, active user communities and detailed documentation can help you get up to speed.

- » **They depend on data democratization.** Projects like OpenStreetMap highlight the collaborative power of open data, enabling anyone to contribute and use high-quality geospatial datasets.

Start with tools like QGIS and GeoServer and pair them with open datasets like Natural Earth or OpenStreetMap to kickstart your GIS journey. See Chapter 21 for more data options.

Spatial Digital Twins

Imagine a digital replica of your city that updates in real time as new data stream in. That's the power of *spatial digital twins*, which are virtual models of physical spaces that let you simulate and analyze real-world scenarios. They're already being used for smart city planning, infrastructure management, and even ecosystem monitoring.

What you need to know about using spatial digital twins:

- » **Real-time data is key.** A digital twin is only as good as the data feeding it. If you're managing a digital twin, make sure that your data sources — such as sensors (devices that collect information from the real world), application programming interfaces (APIs), and other data inputs — are accurate and up to date.

- » **Collaboration is crucial.** Building and maintaining digital twins often requires input from GIS professionals, data scientists, and industry experts. It's a team effort!

- » **Digital twins are resource intensive.** Creating detailed, dynamic digital twins can demand significant computing power, storage, and expertise. If you're new to digital twins, start small by modeling a specific infrastructure system or neighborhood before scaling up to larger areas.

Consider integrating historical data alongside real-time updates in your digital twin. Comparing past and present conditions can help reveal trends, predict future scenarios, and improve decision-making.

Advances in Location-Based Services

From finding the nearest coffee shop to navigating a self-driving car, the use of a location-based service (LBS) is happening everywhere. Advances in LBS are delivering greater accuracy, context-aware services, and new applications in fields like logistics and automated vehicles.

Things to keep in mind about LBS:

- **Privacy matters:** Location-sharing apps and devices often collect user data, so always review and disclose how location data will be stored or used in your projects.
- **Real-world examples:** Geofencing (creating virtual boundaries to trigger actions based on location) helps businesses target customers with location-specific ads, and precision GPS powers drone deliveries and autonomous vehicle navigation.

TIP

Explore geofencing tools to create local marketing campaigns or event notifications.

Citizen Science and Crowdsourcing

Ordinary people are making extraordinary contributions to GIS through crowd-sourced mapping and citizen science projects. Whether they involve disaster response mapping or biodiversity surveys, these efforts generate valuable data that would be impossible for a single organization to collect alone.

Here are some key points to know if you're using crowd-sourced data:

- **Data quality challenges:** Crowdsourced data can vary in accuracy and consistency. Verification systems like peer reviews or cross-referencing with authoritative datasets can help.
- **Empowerment through technology:** Platforms like OpenStreetMap and apps like iNaturalist make it easy for anyone to contribute to GIS projects, fostering inclusivity and global collaboration.

TIP

Join a citizen science initiative, such as iNaturalist, to both contribute and learn. Many projects offer tutorials and support to get you up and running.

CHAPTER 22 **Ten GIS Trends to Watch** 343

Geospatial Education and Workforce Development

As GIS continues to grow, so does the demand for skilled professionals. From university programs to online tutorials, geospatial education is expanding to meet industry demands. Whether you're a beginner or a seasoned pro, resources are available to build your skills. (See Chapter 19 for tips on building a career in GIS).

What you should know about working in the GIS field:

- » **Certifications and hands-on experience matter.** As more and more folks enter the exciting field of GIS, certifications like the GIS Professional (GISP) from the GIS Certification Institute (GISCI) or the technical certifications of Esri can, along with real-world project experience, give you a leg up on the competition.

- » **Stay current.** GIS technologies evolve rapidly, so make lifelong learning part of your career plan. Start with the Esri Academy (https://www.esri.com/training/catalog/search/) for training, including free Massive Open Online Courses (MOOC), or explore OpenGeoHub (https://opengeohub.org/) for learning open-source GIS tools like QGIS, GRASS GIS, and PostGIS.

If you're just starting out, try the many free tutorials in the QGIS Training Manual (https://docs.qgis.org) and Esri to get hands-on experience with real datasets.

Index

A

academia and research, as career path in GIS, 316
accumulation information, 220
accuracy, as factor in choosing data sources, 137
active sensors, 106
adaptive sampling, 109
addresses, turning of into points, 110
adjacency
  analysis of as example of proximity search, 162
  as component of topology, 78
  relative measurement of, 175–176
  as spatial factor, 18
Adobe Color, 272
advanced filtering, as advanced SQL function, 142
AequilibraE plug-in (QGIS), 220, 232
aggregation, as advanced SQL function, 142
A-GPS (assisted GPS), 104
agriculture, use of polygon metrics in, 156
AI (artificial intelligence). See artificial intelligence (AI)
AllTrails, 103
alpha index, 223
Amazon Web Services (AWS), 98, 100
American Society for Photogrammetry and Remote Sensing (ASPRS), 313
analog maps, 59
analysis
  density analysis, 52
  direction analysis, 52
  network analysis, 52, 231
  recognizing spatial nature of, 17–18
  role of GIS in, 10
  service-area analysis, 229–231
  shortest-path analysis, 227
  sinuosity calculation, 52
  suitability analysis, 246
  using GIS for, 49–50

Analysis tab (ArcGIS Pro), 235
Analysis Tools toolbox (ArcGIS Pro), 235
anisotropy, navigation of, 179–180
annual stream pattern, 47
Apple Maps, 103, 109
application developers, role of on internal teams, 304
application programming interfaces (APIs)
  accessing data with, 112–114
  as compared to web service, 296
  getting data through, 113, 297–298
  to many modern GIS data sources, 330
  role of, 112–113
apps, developing interactive ones, 283–287
AR (augmented reality), 3D GIS and, 338–339
Arcade (Esri), 294–295
ArcGIS, 74, 80, 91, 100, 116
ArcGIS Dashboards, 149, 285
ArcGIS Experience Builder, 286
ArcGIS Online, 9, 81, 283, 331, 340
ArcGIS Pro, 76, 92, 140, 141, 143, 147, 154, 155, 158, 161, 163, 190, 194, 197, 205, 210, 212, 215, 219, 220, 221, 222, 231, 232, 234–235, 236, 248, 249, 253, 258, 260, 262, 319, 320
area definition, containment a,k,a., 78
areas. See also polygons
  applying symbology to, 42
  exploration of in raster datasets, 125–126
  how they fit into raster grid system, 71
  illustration of, 64
  a.k.a. polygons, 65
  nearest neighbor analysis as effective tool for, 52
  as noncomparative feature, 41
  as one of three basic shapes of vector data, 134
  in raster data, 64, 65
  as type of space found on maps, 38, 39, 40
  in vector data, 65, 66

artificial intelligence (AI)
  in evolution of geospatial data, 86
  integration of into GIS, 99, 337–338
  use of for searches, 128
ASPRS (American Society for Photogrammetry and Remote Sensing), 313
assisted GPS (A-GPS), 104
attribute data
  evolution of geospatial data, 86
  keeping track of, 85–100
  knowing importance of in GIS, 86–87
attribute index (.atx), 79
attribute searches, 153–154
attribute selection tools, 141
attribute table (.dbf), 75, 79, 140, 215
attribute-based queries, 127
attributes
  examples of, 81
  exploring pixel-level attributes, 92–94
  impedance attributes, 219
  linking objects with, 75–76
  management of with spatial databases, 91–92
  searching by, 160
  using SQL queries based on, 139–141
AudoCAD Map 3D Toolset, 323–324
augmented reality (AR), 3D GIS and, 338–339
Autodesk, 324
automation, of GIS, 293–301
AWS (Amazon Web Services), 98, 100
axis lengths (shape descriptor), 156
Azimuthal Equidistant, 34
azimuthal map projection, 33

B

bar charts, 148
basins
  identifying edges of, 209
  mapping of, 203–204
  working with in GIS, 204
behaviors, defined, 81
Bentley Systems, 327
Berry, Joseph K., 246

binning, as advanced classification technique, 278–279
Blue Marble Geographics, 327
Bluesky, 314
borders, as essential map element, 280
buffering, 243
buffers
  defined, 146, 162
  finding features based on, 146–147
  role of, 162
business
  as career path in GIS, 316
  role of GIS in, 10
Business Analyst extension (ArcGIS Pro), 232

C

calibrated route feature, 111
California Consumer Privacy Act (CCPA), 341
Canada Geographic Information System, 319
Canonical Correlation Analysis (CCA), 260
career development
  exploring career paths in GIS, 315–316
  GIS in, 311–316
  learning and growing in your GIS career, 312–313
  networking and professional growth, 313–315
  understanding what it takes to succeed in GIS, 311–312
cartographer, 27
cartographic design
  for accessibility, 273–275
  color theory, 271–272
  design process, 270–271
  exploring, 270–275
  fonts, 272–273
cartographic modeling
  building models, 260–263
  creating models, 246–247
  defined, 246
  exercising control in, 249
  origins of, 246
  testing models, 264–266

Cassandra, 97
CCA (Canonical Correlation Analysis), 260
CCPA (California Consumer Privacy Act), 341
Census Business Builder, 161
centripetal pattern, 47, 48
certification
   professional certification, 313, 314–315, 344
   technical certification, 314, 344
charts, 148
Cheat Sheet, 3
CHECK constraints, PostGIS/PostgreSQL, 92
circuitry, defining, 223–224
circuits, measuring and modeling of, 223–224
citizen science, 343
CityEngine (Esri), 339
classes, creating of, 276–278
classification, advanced techniques for, 278–279
climate modeling, use of multivariate functions in, 260
clip overlays, 239–240
closed circuit, open network versus, 223
cloud computing/cloud storage
   in evolution of geospatial data, 86
   as revolutionizing GIS, 98
cloud-based GIS, 339–340
clustered sampling, 108, 109
clustering, as advanced classification technique, 278
CNNs (convolutional neural networks), 128
code, reuse of in programming, 299
code of ethics, 315
code page (.cpg), 79
collaboration
   as GIS superpower, 305
   as key to fully understanding patterns and their implications, 53
Color Theory For Dummies (Hibit), 271
color theory (for maps), 271–272
ColorBrewer, 272, 273
comparisons of kind, 41
compliance, as best practice for keeping GIS secure, 100
composite keys (in relational databases), 88
computer hardware, as GIS component, 11–12
computers, helping them read maps, 58
conceptual models
   as compared to mental blueprint, 58
   creation of, 57–68
conic map projection, 33, 34
connectivity
   as component of topology, 78
   measurement of, 213–217
   as spatial factor, 18
   why it matters, 214
conservation
   as career path in GIS, 316
   role of GIS in, 10
   soil conservation, 207
   use of cartographic models in, 247
containment
   as component of topology, 78
   relative measurement of, 177–178
Content Standard for Digital Geospatial Metadata (CSDGM), 115–116
contiguity, adjacency a.k.a., 78
continuous data, 191
continuous surfaces, 185, 186, 188
contour lines, 67
control access, as best practice for keeping GIS secure, 100
convolutional neural networks (CNNs), 128
coordinate systems, 34–35
coordinates, finding objects by, 72–74
Copernicus Data Space Ecosystem, 332–333, 340
Create Fishnet (ArcGIS Pro), 190
credits, as essential map element, 280
crowdsourcing, 343
CRS (geographic coordinate reference system), 31, 32
"crystal ball" approach, to keeping on track, 138–139
CSDGM (Content Standard for Digital Geospatial Metadata), 115–116
CSV, as common GIS data format, 14
custom breaks, as advanced classification technique, 278
cylindrical map projection, 33–34

D

D8 method (ArcGIS Pro and QGIS), 205
dashboards, 148, 284–285
data
    accessing of with API connections, 112–114
    anonymization of as best practice for keeping GIS secure, 100
    breaking down of to define map's contents, 61–63
    building data about, 114–117
    choice of to define map's contents, 59–60
    collecting of with GPS receiver, 103–105
    common GIS data formats, 14–15
    comparison of using interval and ratio scales, 41–42
    connecting of with joins and relationships, 90–91
    continuous data, 191
    deciding whether value-added data is worth the cost, 335
    evaluating factors for collecting high-quality data, 102–103
    geocoding of, 109–112
    geographic data. *See* geographic data
    geospatial data, evolution of, 86
    getting it through APIs and webservices, 297–298
    GIS data sources, 329–335
    grouping and ranking of, 158–159
    identifying quality data, 101–103
    importing, 103–109
    layering of, 234
    mapping data, 275–279
    matching symbols with data types, 40–42
    selection of from spatial data layers, 62
    summarizing, 128–131
    in thematic maps, 28
    types of. *See* raster data; vector data
    use of remote sensing for collection of, 106
    verifying characteristics of to define map's contents, 63
    working with samples of, 108–109
data dashboards, 148, 284–285

data encryption, as best practice for keeping GIS secure, 100
data exports, automating, 287
data integrity tools, 92
data interoperability, incorporating data interoperability, standards, and security, 99–100
data management, role of GIS in, 9
Data Management toolbox (ArcGIS Pro), 258
Data Management Tools (QGIS), 235
data models, understanding GIS data models, 69–84
data security, incorporating data interoperability, standards, and security, 99–100
data standards, incorporating data interoperability, standards, and security, 99–100
data storage, emerging systems for, 97. *See also* cloud computing/cloud storage
data structure, as affecting retrieval, 135–136
data type. *See also* raster data; vector data
    as factor in choosing data sources, 136–137
    sample data use cases, 138
database formats, 14
database management systems (DBMs)
    searching with, 127–128
    working with, 87–95
datum, as component of reference system, 31, 32
decision makers, role of on internal teams, 304
default cutoff value, 219–220
dendritic pattern, 47, 48
denominator, in map's scale, 27
density, as spatial factor, 17
density analysis, 52
Density-based Clustering tool (ArcGIS Pro), 163
Density-Based Spatial Clustering of Applications with Noise (DBSCAN) (QGIS), 163
depressionless surface, 204
design, cartographic. *See* cartographic design
diagonal paths, 172, 173
Digital Elevation Model (DEM)
    for calculating viewsheds, 202
    conversion of into different formats, 83

defined, 122
illustration of, 67, 123
for locating streams, 209
ModelBuilder (ArcGIS Pro) as including, 263
for modeling flow direction, 205, 208
as raster-based surface model, 81
sizes of, 82–83

digital maps, 59, 63–68, 282

digital twins, 12

dimension, representation of when everything is square, 71–72

direction analysis, 52

disaster management
role of GIS in, 10
use of polygon metrics in, 156

discrete surfaces, 185, 186

disease breakouts, study of using polygon metrics, 158

distance
calculating of along networks, 174
finding features based on, 146–147
functional distance, 178–182
measurement of, 169–182

distance-weighted interpolation, 195–196

distributions
defined, 152
exploration of in raster datasets, 125–126

diverse features, seeing patterns among, 45

documentation, of programming process, 300–301

domains, in geodatabases, 91

dynamic classification, as advanced classification technique, 279

E

Earth
flattening of, 29–36
measurement of, 29–32

EarthExplorer (USGS), 333

Eason, Ken, 309–310

economic geography, example of, 59

edges, as one of three key components of network datasets, 225

education, as best practice for keeping GIS secure, 100

elements, of maps, 58

elevation, GIS data models as representing, 81

elevation grid, 204

ellipsoid, 30, 31

emergency response planning
ideal data type for, 138
using polygon metrics, 158

empty results, 243

encoding, and using turns and intersections, 225–226

Enterprise, 81

environmental management, use of multivariate functions in, 260

environmental planning, role of connectivity in, 214

environmental science and conservation
as career path in GIS, 316
role of GIS in, 10

environmental studies, use of polygon metrics in, 156

Eratosthenes, 29

ERDAS IMAGINE, 325–326

errors, handling of in programming, 299–300

Esri (Environmental Systems Research Institute), 9, 81, 88, 92, 287, 296, 314, 320, 331, 338, 339

Esri Academy, 344

Esri REST API, 298

ethics, evolving geospatial data privacy and ethics, 341

Euclidean distance, 170

Excel, as common GIS data format, 14

Experience Builder, 149

Extract Nodes (QGIS), 215

F

fastest path, finding, 227–228

Feature to Point tool (ArcGIS Pro), 215

Index   349

features
- analyzing linear feature, 124–125
- finding of in vector data, 133–150
- identifying and locating of in raster data, 121–126
- identity feature, 238
- line features, 41
- linear features, 45–46, 124–125
- locating map features, 159–164
- locating specific features of vector data with SQL, 139

Federal Geographic Data Committee (FGDC), Content Standard for Digital Geospatial Metadata (CSDGM), 115–116
feedback, incorporation of, 286–287
Felt, 283
Field Calculator (QGIS), 228
File Geodatabase (.gdb), 143
filters, in interactive maps, 284
Find Best Route (ArcGIS Pro), 228
First Law of Geography, Tobler's, 195
first-order streams, 210–211
5G networks, use of ray tracing in, 201
Fix Topology Error tool (ArcGIS Pro), 243
flood prediction, use of flow direction in, 207
flow direction
- modeling of in GIS, 205–206, 207–208
- why it matters, 205

flow speed, determining, 206
flowcharts, use of, 59, 62
flow-direction analysis, 204
focal functions, 251, 254
Focal Statistics raster function (ArcGIS Pro), 253, 254
fonts (in maps), 272–273
foreign keys (in relational databases), 88, 89
friction, GIS's assigning of, 181
friction surface, 181
functional distance, 178–182
functional surface, 181
functions
- focal functions, 251, 254
- global functions, 258–260
- hydrology functions, 259
- interpolation functions, 258
- local functions, 249–251
- multivariate functions, 259
- neighborhood functions, 251–254
- surface functions, 258
- zonal functions, 254–257

G

Galileo (European satellite), 104
gamma index, 213, 214–217
Gaussian curves, 174
GCS (geographic coordinate system), 34, 172
General Data Protection Regulation (GDPR), 341
Generate Random Points (ArcGIS Pro), 190
GeoAI (Geospatial Artificial Intelligence), 337–338
geocoder, role of, 110
geocoding, of data, 109–112
geodatabases
- as advanced data model, 78
- illustration of land use geodatabase layer with associated attribute table, 92
- leveraging of for success, 98–99
- as organizing data into objects, 97
- role of, 80, 81
- as spatial database format, 91

geodesy, 29
Geodetic Reference System of 1980 (GRS80), 30
geographic coordinate reference system (CRS), 31, 32
geographic coordinate system (GCS), 34, 172
geographic data
- collecting, 101–117
- as GIS component, 11, 13–15
- identifying quality data, 101–103
- primary data, 13
- secondary data, 14

geographic software, as GIS component, 11, 12–13
geography
- searching vector data with, 144–147
- why it matters, 16–18

geoid, 30, 31
GeoJSON, 14, 100, 143
geometry, understanding geometry and patterns in, 48–49
GeoPackage, 80, 91, 100, 143, 144
GeoPlatform.gov, 334
GeoServer, 9, 325, 340, 342
geospatial architects, 16
Geospatial Artificial Intelligence (GeoAI), 337–338
geospatial data, evolution of, 86
geospatial database, defined, 87
geospatial education, 344
The Geospatial Professional Network (GPN, formerly URISA), 313, 314
geospatial workforce development, 344
Geostatistical Analyst extension (ArcGIS Pro), 197
geostatistics and machine learning, 15
GIS
  automating, 293–301
  best practices for keeping secure, 100
  building career in, 311–316
  as cloud-based, 8–9
  core concepts of, 19–22
  defined, 11
  integrating AI and machine learning in, 99
  networking options in, 227
  origins of, 8
  power of, 8–10
  purpose of, 16
  as transforming data into action, 9–10
  trends in, 337–344
  understanding maps in, 22–29
  use of for analysis, 49–50
  vector data as drawing tools of, 134
GIS administrators, 16
GIS analysts and consultants, 16, 304
GIS Certification Institute (GISCI), 313, 344
GIS collective, 11–16
GIS data models, 69–84
GIS Professional (GISP) certification, 313, 315, 344
GIS projects, starting at the end as best, 59, 60, 61

GIS software applications, 74, 75, 76, 81, 83, 87–88, 91, 93, 94, 140, 154, 206, 211, 319–327. *See also specific applications*
GIS technicians, role of on internal teams, 304
GISCI Geospatial Core Technical Knowledge Exam, 315
GIS-Pro (GPN), 314
global functions, 258–260
Global Mapper, 326–327
Global Navigation Satellite System (GNSS), 104
GLONASS (Russian satellite), 104
GNSS base station, 104
Google AppSheet, 286
Google Cloud, 98, 100
Google Earth, 143
Google Earth Engine, 127, 324, 338
Google Maps, 103, 109, 143, 178, 226, 284, 296, 298
government, as career path in GIS, 315
GPN, formerly URISA (The Geospatial Professional Network), 313, 314
GPS technology
  collecting data with GPS receiver, 103–105
  in evolution of geospatial data, 86
  typical handheld GPS device, 105
graduated, 41
Grafana, 285
graphic design, of maps, 280–282
GRASS GIS, 9, 197, 259, 260, 322–323
graticule, as essential map element, 280
great circle distance, 171–172
grid cell resolution, 72
grid cells
  defined, 71
  finding of by category, 74
  measuring distances in, 172
  quirks in, 205–206
grid coordinates, 35
grid overlay, 70
Grid Vector Layer (QGIS), 190

grids
  numbering in, 72–73
  searching of in raster data model, 74
groups/grouping
  by common properties, 164–165
  defining ones you want to find, 164
  by location and patterns, 165–166
  by what you already know, 166
GRS80 (GCS), 172
GRS80 (Geodetic Reference System of 1980), 30

H

hazardous material maps, 25
head gradient, 259
heat maps, 148
hexbin mapping, as advanced classification technique, 278–279
hierarchy
  as characteristic of objects, 96
  for traffic, 220
high friction, 181
higher-order streams, 211
HOT Export Tool, 116
hot-spot analysis, 157, 158
HTTPS, use of as best practice for keeping GIS secure, 100
human surfaces, 184
hydrology functions, 259
hypotenuse, 171

I

icons, explained, 2–3
identity feature, 238
identity overlays, 238–239
IDW (inverse distance weighting), 195, 197
Image Analyst extension (ArcGIS Pro), 253
image enhancement, 107
images, enhancing and classifying of, 107
impedance, key impedance layer settings, 219–220
impedance values, working with, 217–221

industry models, leveraging of for success, 98–99
INEGI (National Institute of Statistics, Geography, and Data Processing) (Mexico), 334–335
infrared, in remote sensing, 106
infrastructure design, use of flow direction in, 207
infrastructure resilience, role of connectivity in, 214
inheritance, as characteristic of objects, 96
inner join, 90
intangibles, consideration of in functional distance, 180–181
interactive web maps, 283–287
International Organization for Standardization (ISO), ISO 19115, 116
International User Conference (Esri), 314
Internet of Things (IoT), in evolution of geospatial data, 86
interoperability, ensuring of with standards, 99–100
interpolation
  cheat sheet for, 197–198
  defined, 189, 191
  distance-weighted interpolation, 195–196
  kriging, 196, 197
  linear interpolation, 192–194, 197
  natural neighbor, 198
  nearest neighbor, 198
  non-linear interpolation, 192
  predicting values with, 191–198
  spline interpolation, 198
  T intersection, 226
  trend surface analysis, 196, 197
interpolation functions, 258
Interpolation toolset (QGIS), 197
intersect, as spatial operator, 145, 146
Intersect tool (ArcGIS Pro), 237
Intersect tool (QGIS), 237
Intersection (QGIS), 237
intersection overlays, 237–238
intersections, 224–226
interval scales, 39, 40, 41–42, 188
inverse distance weighting (IDW), 195, 197

IoT (Internet of Things), in evolution of geospatial data, 86
IPUMS National Historic GIS (NHGIS), 332
irregular, as component TIN model breaks surfaces down into, 83
ISO 19115, 116
Isochrones (QGIS), 232
isolation
　relative measurement of, 177
　as spatial factor, 18
isotropy, defined, 179
IT specialists, 16

J

JavaScript, 293, 294, 295
Join Attributed by Location tool (QGIS), 235
joins
　as advanced SQL function, 142
　connecting data with, 90–91
　types of, 90
JPEG2000, as common GIS data format, 14
junctions, as one of three key components of network datasets, 225

K

Kepler.gl, 284
kernel-density estimation, 157, 158
K-Means Clustering, 260
KML (.kml), 143, 144
kriging, 196, 197

L

labels, as essential map element, 280
Lambert Conformal Conic, 34
land boundaries, defining, 35
land development studies, ideal data type for, 138
land-use tracking, use of GIS for, 8
latitude, 31, 32, 34
law enforcement maps, 25

layers
　in interactive maps, 284
　OpenLayers, 9, 286
　raster data, 74–75
　selection of data from spatial data layers, 62
　working with map layers, 74–75
LBSs (location-based services), advances in, 343
Leaflet, 9, 286
left join, 90
legend, as essential map element, 279
libraries (in programming), 296
Light Detection and Ranging (LiDAR), 189
line features, 41
linear features
　analyzing of, 124–125
　describing patterns with, 45–46
linear interpolation, 192–194, 197
linear referencing system (LRS), 111–112
line-of-sight, simulating of with ray tracing, 201
lines
　applying symbology to, 42
　creation of by reference, 111
　how they fit into raster grid system, 71
　as ideal data type for transportation analysis, 138
　as ideal data type for water resource management, 138
　illustration of, 64
　nearest neighbor analysis as effective tool for, 52
　as one of three basic shapes of vector data, 134
　in raster data, 65
　as type of space found on maps, 38, 39, 40
　in vector data, 65, 66
LinkedIn, 314
links, 214–215
lists
　attribute-based lists, 288
　hierarchical lists, 288
　spatially defined lists, 288
Living Atlas of the World, 331
local functions, 249–251

location, selecting by, 234–235
location-based services (LBSs), advances in, 343
longitude, 32
low friction, 181
LRS (linear referencing system), 111–112

M

M values (measure values), 111
machine learning (ML)
  in evolution of geospatial data, 86
  integration of into GIS, 99, 337–338
  use of for searches, 128
Manhattan distance, measurement of, 172–174
many-to-many (relationship supported by), 91
map algebra
  defined, 130, 247
  exercising control in cartographic model, 249
  performing functions in, 248–260
  uses of, 247–248
  using local functions, 249–251
Map Analysis Package (MAP), 74
map layers, working with, 74–75
map overlay
  advanced overlay techniques, 236–241
  basic overlay methods, 234–235
  raster overlay, 242–244
  troubleshooting overlay issues, 243–244
map projections, 32–35
Mapbox, 9, 284
MapColPal, 272
MapInfo Pro, 321–322
mapped areas, as essential map element, 280
mapping, role of GIS in, 10
mapping data
  creating classes, 276–278
  qualitative data, 275–276
  quantitative data, 276
MapQuest, 109
maps
  analyzing and quantifying patterns, 48–52
  categorizing space on, 38–39
  conversion of into digital formats, 63–68
  data in thematic maps, 28
  defining contents of, 59–63
  developing interactive ones, 283–287
  digital maps, 59, 63–68, 282
  essential elements, 279–280
  getting scale of right, 27–29
  graphic design of, 280–282
  helping computers read maps, 58
  interactive features of, 284
  interpreting results and making decisions, 53
  interpreting symbols on, 26–27
  laying out yours, 279–283
  locating features of, 159–164
  making sense of symbols on, 38–42
  mapping nearby parks for your neighborhood, 116–117
  optimizing for different media, 282–283
  printed maps, 283
  as questions as well as answers, 36
  reading, analyzing, and interpreting, 37–53
  recognizing patterns in, 42–48
  as representing spatial data, 22
  topographic maps, 19, 23, 25–26
  types of, 22–23
  understanding geometry and patterns in, 48–49
  understanding of in GIS, 22–28
Massive Open Online Courses (MOOCs), 344
maximum function, 251, 252
Maxwell, John C., 308
mean (average), as type of simple statistic, 129
measurement
  of distance along networks, 174
  of distances, 169–182
  of distances in grid cells, 172
  on flat surface, 171
  of functional distance, 178–182
  of Manhattan distance, 172–174
  relative measurements, 175–178
  on spherical Earth, 171–172
  taking absolute measurement, 170–174
  understanding levels of, 39–40

media, optimizing maps for different media, 282–283
median, as type of simple statistic, 129
medical geography, example of, 59
medical maps, 24–25
Mercator projection, 33–34
meridians, 32
metadata
  as best friend, 150, 330
  creation of, 115–116
  defined, 15, 115
  file extension (.xml), 79
  role of, 115
methodology, application of in defining contents of maps, 60–61
methods, as GIS component, 11, 15–16
Microsoft Azure, 98, 100
microwave, in remote sensing, 106
military, use of viewsheds in, 200
minimum/maximum, as type of simple statistic, 129
ML (machine learning). *See* machine learning (ML)
mode, as type of simple statistic, 129
model building
  defined, 260
  implementation of model, 262–263
  planning your analysis, 261–262
  preparing data inputs, 262
  testing models, 264–266
Model Designer (QGIS), 262
ModelBuilder (ArcGIS Pro), 262, 263
model-creation process, embracing, 58–59
MongoDB, 97
MOOCs (Massive Open Online Courses), 344
movement, as spatial factor, 18
multi-band rasters, 93
multidimensional rasters, 93
multivariate functions, 259

N

NaCIS (North American Cartographic Information Society), 313
NAD83 (GCS), 172, 244

NASA
  Earthdata API, 298
  SEDAC (Socioeconomic Data and Applications Center), 333
National Institute of Statistics, Geography, and Data Processing (INEGI) (Mexico), 334–335
Natural Earth, 331–332, 342
natural neighbor, 198
Natural Neighbor (ArcGIS Pro), 197
natural resource management
  use of multivariate functions in, 260
  use of viewsheds in, 200
near, as spatial operator, 145, 146
nearest neighbor, 198
nearest neighbor statistic, 50, 51, 52
nearness, relative measurement of, 175–176
neighborhood functions, 251–254
network, as component TIN model breaks surfaces down into, 83
network analysis, 15, 52, 231
Network Analysis toolbox (QGIS), 219
Network Analyst extension (ArcGIS Pro), 219, 221, 228, 232
Network Dataset Properties (ArcGIS Pro), 228
network datasets, key components of, 225
networks
  defining circuitry, 223–224
  directing traffic and exploiting networks, 226–231
  measuring connectivity, 213–217
  navigating one-way paths, 221–222
  non-planar network, 216
  planar network, 216
  working with impedance values, 217–221
  working with in GIS, 231–232
  working with turns and intersections, 224–226
nicest path, finding, 228–229
nodes, 78
nominal features, working with, 41
nominal scales, 39, 40, 41
noncartographic outputs, 287–288
noncomparative features, 41
non-linear interpolation, 192, 194–195

Index   355

non-planar network, 216
non-uniform surfaces (anisotropy), 179–180
North American Cartographic Information Society (NaCIS), 313
north arrow, as essential map element, 279–280
NorthStar of GIS, 313
NoSQL (a.k.a. "not only SQL") databases, 95–99
numerator, in map's scale, 27

O

object-oriented database management system (OODBMS), 95–99
object-oriented models, building smarter systems with, 81
object-oriented programming (OOP), 81
objects
  attribute searches, 153–154
  geodatabases as organizing data into, 97
  as powerful for GIS, 96
  searching for right ones, 153–159
observer height, as factor in making viewsheds more realistic, 202
obstructions, as factor in making viewsheds more realistic, 202
OD (Origin-Destination) matrix, 220–221
one-dimensional, 40
one-to-many (relationship supported by GIS), 91
one-to-one (relationship supported by GIS), 91
one-way paths, 221–222
one-way systems, understanding of, 221–222
online communities, 314
OODBMS (object-oriented database management system), 95–99
OOP (object-oriented programming), 81
Open Geospatial Consortium (OGC), 99–100
open network, versus closed circuit, 223
OpenCities Map, 327
OpenGeoHub, 344
OpenLayers, 9, 286
open-source GIS and data, growth of, 341–342
open-source GIS solutions, evolution of, 9
OpenStreetMap (OSM), 9, 21, 116, 330–331, 341, 342
OpenStreetMap Overpass API, 298
OpenWeatherMap, 112, 113–114, 298
Optimized Hot Spot Analysis tool (ArcGIS Pro), 158
ordinal features, depiction of, 41
ordinal scales, 39, 40, 41
organizations
  adapting GIS for small organizations, 306
  adapting GIS to needs of, 305–306
  aligning GIS with goals of, 307
  designing and implementing GIS in your organization, 307–311
  external partners, 305
  GIS in, 304–311
  GIS roles and interactions in, 304–305
  internal teams, 304
  managing people problems, 308–309
  overcoming integration challenges, 308
  planning for integration, 309–311
  transformation of with GIS, 304–306
orientation (shape descriptor), 154
orientation, determining, 187–188
Origin-Destination (OD) matrix, 220–221
Origin-Destination Cost Matrix (ArcGIS Pro), 219
orthogonal paths, 172, 173
OSM (OpenStreetMap). *See* OpenStreetMap (OSM)
outer join, 90
outputs
  cartographic outputs, 270–287
  generating of with GIS, 269–288
  noncartographic outputs, 287–288
overlay analysis, 15
overlay methods, exploring basic ones, 234–235
overlay techniques, advanced, 236–241
Overlay toolset (ArcGIS Pro), 235
ownership, defining, 35

P

pair-wise, 51
pair-wise nearest neighbors, 51
pan, in interactive maps, 284
parallel pattern, 47, 48

parallels, 31, 34
passive sensors, 106
path-based, in viewshed analysis, 200
paths
  fastest path, 227–228
  nicest path, 228–229
  shortest path, 227
pattern change, as spatial factor, 18
patterns
  analyzing and quantifying, 48–52
  analyzing patterns in real world, 49
  clustered patterns, 43–44
  describing patterns with linear features, 45–46
  determining types of, 50–52
  figuring out why they exist, 53
  random patterns, 43, 51
  recognizing, 42–48
  repeated sequence of shapes, 46–47
  seeing patterns among diverse features, 45
  understanding geometry and patterns in, 48–49
  uniform patterns, 44
PCA (Principal Component Analysis), 260
people, as GIS component, 11, 16
performance, optimization of in programming, 300
Personally Identifiable Information (PII), 340
PHI (Protected Health Information), 340
physical geography, example of, 59
physical surfaces, 184
pie charts, 148
pixel grid, 122, 124
pixels, 20
planar map projection, 33, 34
planar network, 216
PLSS (U.S. Public Land Survey System), 35
points
  analyzing distributions of, 157
  creation of by reference, 111
  how they fit into raster grid system, 71
  as ideal data type for emergency response planning, 138
  illustration of, 64
  nearest neighbor analysis as effective tool for, 52
  as one of three basic shapes of vector data, 134
  turning addresses into, 110
  as type of space found on maps, 38, 39, 40
  in vector data, 64, 65, 66
point-to-point, in viewshed analysis, 200
polygons
  areas a.k.a, 65
  as distinct from points and lines, 152
  as ideal data type for land development studies, 138
  as ideal data type for wildlife habitat mapping, 138
  metrics of, 154–156
  searching of in a GIS, 152–153
  sliver polygons, 237, 243
pop-ups, in interactive maps, 284
PostGIS, 9, 78, 88, 127
PostGIS/PostgreSQL, CHECK constraints, 92
PostgreSQL, 78, 127
PostgreSQL/PostGIS, 81
pour points, 204, 208, 209
Power BI with Map Visuals (Microsoft), 285
PowerBI, 149
precision, as factor in choosing data sources, 137
primary keys (in relational database), 88, 89
prime meridian, 32
Principal Component Analysis (PCA), 260
printed maps, 283
privacy
  addressing of in GIS, 100
  evolving geospatial data privacy and ethics, 341
Processing Toolbox (QGIS), 197, 228, 235, 258
professional certification, 313, 314–315
professional organizations, 313
programming best practices, 298–301
programming jargon, 295–296
programming languages, 293–295
Project (ArcGID Pro), 244
projected coordinate system, 34
projection (.prj), 79
Protected Health Information (PHI), 340
proximity searches, 162

public health
  role of GIS in, 10
  use of cartographic models in, 247
Pythagoras, 171
Pythagorean Theorem, 171
Python language, 113, 114, 149, 293, 294, 295, 296

Q

QGIS (formerly Quantum GIS), 9, 74, 76, 81, 91, 92, 100, 116–117, 124, 141, 143, 144, 147, 154, 161, 163, 190, 194, 197, 205, 210, 212, 215, 219, 220, 221, 222, 228, 231, 232, 235, 236, 248, 249, 251, 253, 258, 260, 262, 319, 320–321, 338, 341, 342
QGIS Expressions, 295
QGIS Training Manual, 344
QGIS2Web, 284
QNEAT3 plug-in (QGIS), 220, 221, 228, 232
qualitative data, mapping of, 275–276
quality, understanding why it matters, 102
Query Builder (ArcGIS Pro), 155

R

R language, 113, 114, 149, 294
radial pattern, 47, 48
Random Points in Polygon (QGIS), 190
random sampling, 109
Raster Calculator (QGIS), 124
Raster Calculator tool (ArcGIS Pro), 210, 248
raster cells, 93
raster data
  benefits of versus vector data, 66–68
  calculating distance along networks, 174
  as compared to vector data, 93–94, 133
  defined, 20, 63
  exploring world through, 121–131
  identifying and locating features in, 121–126
  locating areas of interest, 122–124
  as made of grid of pixels, 93
  performing searches in, 126–128
  resampling, 257–260
  visualizing and interpreting of, 130–131
raster data models
  examination of, 70–76
  finding grid cells by category, 74
  finding objects by coordinates, 72–74
  limitations of, 82
  linking objects with attributes, 75–76
  making quality difference with resolution, 72
  representation of dimension when everything is square, 71–72
  storing surface data in, 82
  as working well when overview is needed, 84
  working with map layers, 74–75
raster grid cells, 72
raster grid coordinate system, 73
raster grids, 70
raster images, as common GIS data format, 14
raster layers, 74–75
raster queries, 93
raster-based model, representation of data in, 64–65
ratio scales, 39, 40, 41–42, 188
ray tracing, 201–202
real estate, use of viewsheds in, 200
real-time classification, as advanced classification technique, 279
reclassification, as tool for performing quick searches in rasters, 127
rectangular stream pattern, 47, 48
reference line, 111
reference maps, 19, 23, 24, 27
reference system, for defining and measuring locations on Earth's surface, 30, 31
Regression Analysis, 260
relational data, structuring of, 88–90
relational database management system (RDMS), 75, 87, 88
relational database, naming of, 88
relational tables, role of, 88
relationships
  building of with vector data, 135
  types of supported by GIS, 91
remote sensing
  expansion of capabilities of, 340
  passive versus active remote sensing, 106
  use of to collect data, 106

reports, automating, 287
Reproject Layer (QGIS), 244
Resample tool (ArcGIS Pro), 258
resampling, of raster data, 257–260
resolution, making quality difference with, 72
restrictions (for traffic), 220
results
  empty results, 243
  making the most of search results, 142–144
  managing errors and uncertainty in, 149–150
  reporting and sharing of, 131
  validating and verifying of, 149–150
  visualizing and summarizing data, 148
retail site analysis, using polygon metrics, 158
retrieval, how data structure affects, 135–136
ridge lines, 204
risk assessment, use of cartographic models for, 246
road map, 24
roundness (shape descriptor), 154
route optimization, role of GIS in, 10
rugged surfaces, 186–187

S

SaaS (Software as a Service) solutions, 339–340
SAGA GIS, 259
SAGA NextGen plug-in (QGIS), 260
sampling. *See also* resampling
  methods of, 108–109
  statistical surfaces, 189–190
  stratification of, 109
  working with samples of data, 108–109
sans serif fonts, 273
satellite imagery, in evolution of geospatial data, 86
satellites
  as collecting data in raster format, 106
  network of for GPS to work, 104
scale, as factor in choosing data sources, 137

scale bar, as essential map element, 279
SCP (Semi-Automatic Classification Plugin) (QGIS), 338
scripts/scripting
  fetching data from APIs through, 113
  making GIS work smarter with, 296–297
searches
  attribute searches, 153–154
  by attributes, 160
  combining multiple search methods, 163–164
  for geographic objects, distributions, and groups, 151–166
  by groups and clusters, 162–163
  making the most of search results, 142–144
  optimizing data for speedy searches, 135–136
  performing of in raster data, 126–128
  of polygons in a GIS, 152–153
  by proximity, 162
  for right objects, 153–159
  searching vector data with geography, 144–147
  by shape and size, 160–161
  in simple rasters, 127
  using machine learning and AI for, 128
security, addressing of in GIS, 100
SEDAC (Socioeconomic Data and Applications Center) (NASA), 333
Select by Attributes interface (ArcGIS Pro), 155
Select by Location tool, 234
Selection group of Map tab (ArcGIS Pro), 234
Semi-Automatic Classification Plugin (SCP) (QGIS), 338
separation, relative measurement of, 175, 177
serif fonts, 273
Service Area (ArcGIS Pro), 219
service areas, defining, 229–230
service-area analysis, 229–231
shape, as spatial factor, 18
shape descriptors, 154
shape format (.shp), 79
shape index (.shx), 79

shapefiles
    common files of, 79
    as common GIS data format, 14
    core files of, 79
    defined, 79
    formats of, 143
    limitations of, 80
    use of for easy data exchange, 79
shapes, understanding repeated sequence of, 46–47
shared keys (in relational databases), 88
shelterbelt networks, 215
shortest straight-line path, finding, 170–172
shortest-path analysis, 227
Shreve method, 211, 212
sinuosity
    calculation of, 52
    as spatial factor, 18
site selection, use of cartographic models in, 247
size, as spatial factor, 18
sliver polygons, 237, 243
slope, determining, 187–188
smartphone
    accuracy of, 105
    as GPS receiver, 103–104
smooth surfaces, 186–187
Snap Geometries to Layer tool (QGIS), 243
snapping, 243
The Society of Conservation GIS, 313
Socioeconomic Data and Applications Center (SEDAC) (NASA), 333
software applications (for GIS). *See* GIS software applications
Software as a Service (SaaS) solutions, 339–340
soil conservation, use of flow direction in, 207
solar analysis, use of ray tracing in, 201
space, categorization of on maps, 38–39
spaghetti model, 76
spatial analysis, Tobler's First Law of Geography, 195
Spatial Analyst extension (ArcGIS Pro), 253, 260
spatial data, 20–22
spatial data layers, selection of data from, 62

spatial data science, use of GIS for, 8
spatial databases, managing attributes with, 91–92
spatial digital twins, 342
spatial index for features (.sbn/.sbx), 79
spatial index (.qix), 79
spatial indexing, 135–136
spatial interpolation, 15
spatial join, 90, 142
Spatial Join tool, 234, 235
spatial joins, performing, 235
spatial operators, 145–146
spatial queries, 123
spatial relationships
    defined, 144
    use of relative measurement tools to evaluate, 175–176
spatial sampling methods, 108–109
spatial smarts, adding of to GIS with topology, 77–78
spatial variation, as ingredient of surface, 184
species habitats, mapping of using polygon metrics, 157
spline interpolation, 198
Spline tool (ArcGIS Pro), 194
SQL (Structured Query Language). *See* Structured Query Language (SQL)
SQLite/GeoPackage, 88
standard deviation, as type of simple statistic, 129
standards, ensuring interoperability with, 99–100
State of the Map (OpenStreetMap), 314
statistical surfaces
    examining character of, 183–188
    predicting values with interpolation, 191–198
    sampling, 189–190
    working with surface data, 188–191
statistics
    advanced statistical techniques, 130
    calculation of, 128–131
    counting, tabulation, and summary statistics, 147–149
    mixing statistics and GIS, 259–260
    performing basic statistical analysis, 148

simple statistics, 129–130
summary statistics, 288
types of, 129–130
steep surfaces, 187
Strahler method, 211, 212
straight-through movements, 226
strata, dividing into, 109
Strava, 103
stream networks, 208–212
stream order, 210–211
stream patterns, 47, 48
streams, 208–209
Structured Query Language (SQL)
   as compared to object-oriented and NoSQL, 96
   defined, 94, 154
   exploring advanced functions of, 142
   locating specific features with, 139–144
   searching with, 94–95
   use of, 295
   using SQL queries based on attributes, 139–141
sub-basins, 208, 209
sub-parallel pattern, 47
subtypes, in geodatabases, 92
suitability analysis, use of cartographic models in, 246
supervised classification, 107
surface data, 188–191
surface functions, 258
surfaces
   breaking rules with, 191
   categories of, 184
   choosing right model for, 84
   continuous surfaces, 185, 186, 188
   dealing with, 81–84
   depressionless surface, 204
   discrete surfaces, 185
   friction surface, 181
   functional surface, 181
   human surfaces, 184
   ingredients of, 184
   non-uniform surfaces (anisotropy), 179–180
   physical surfaces, 184
   in raster data, 65
   representing them in vector model, 83–84
   rugged surfaces, 186–187
   smooth surfaces, 186–187
   statistical surfaces, 183–198
   steep surfaces, 187
   storing surface data in raster model, 82
   topographic surfaces, 199–212
   use of term in GIS, 183
   in vector data, 66
   volumes a.k.a., 71
surroundedness, relative measurement of, 177–178
symbols
   design of, 42
   as essential map element, 280
   interpretation of on maps, 26–27
   making sense of, 38–42
   matching of with data types, 40–42
symmetrical difference overlay, 241–242
systematic sampling, 108, 109

T

T intersection, 225, 226
Tableau, 149, 285
tables
   connecting data with joins and relationships, 90–91
   defining relationships between, 90–91
   how joins work in GIS, 90
   structure of relational data, 88–90
   working with, 87–95
tabular, as common GIS data format, 14
Team GIS, players of, 11
Teamwork Makes the Dream Work (Maxwell), 308
technical certification, 314, 344
technology and software development, as career path in GIS, 316

telecommunications, use of viewsheds in, 200
terrain, GIS data models as representing, 81
thematic maps, 19, 23–25, 28, 37
themes, breaking data down into, 58
3D Analyst extension (ArcGIS Pro), 197
3D Tiles, 100
3D GIS, and augmented/virtual reality, 338–339
3D topographic maps, use of DEM for, 82
thresholding, as tool for performing quick searches in rasters, 127
TIFF, as common GIS data format, 14
time sliders, in interactive maps, 284
TIN (Triangulated Irregular Network), 65, 67, 81, 83–84
TIN Interpolation tool (QGIS), 194
title, as essential map element, 279
Tobler's First Law of Geography, 195
Tomlin, C. Dana, 246
Tomlinson, Roger, 8, 319
tools, choosing appropriate ones, 52–53
topographic analysis, role of GIS in, 10
topographic maps, 19, 23, 25–26
topographic surfaces
  characterizing water flow, 205–208
  defining streams, 208–212
  exploring, 199–212
  mapping watersheds and basins, 203–204
  modeling visibility with viewsheds, 200–202
topological model, 77
topology
  adding spatial smarts to GIS with, 77–78
  components of, 78, 135
  defined, 77
  role of, 78, 135
topology validation tools, 243
traffic
  directing of, 226–231
  modeling impedance in traffic networks, 218–219
  understanding why it is good or bad, 218
Transactions table, illustration of, 89

transportation analysis, ideal data type for, 138
transportation geography, example of, 59
transportation studies, use of cartographic models in, 247
Transverse Mercator projection, 34
Travel Attributes section (ArcGIS Pro), 228
trellis stream pattern, 47, 48
trend surface analysis, 196, 197
trends (in GIS), 337–344
triangulated, as component TIN model breaks surfaces down into, 83
Triangulated Irregular Network (TIN), 65, 67, 81, 83–84
triangulation, 29
tributary streams, 208, 209
trunk stream, 208, 209
turn directions, 226
turn impedance, 226
turns, 224–226
two-dimensional, 40
two-dimensional squares, 72

U

unidirectional paths, 221
union overlays, performing, 236–237
Unity, 339
Universal Transverse Mercator (UTM), 34, 35
unsupervised classification, 107
urban design, use of viewsheds in, 200
urban development/urban planning
  as career path in GIS, 316
  role of connectivity in, 214
  role of GIS in, 10
  use of cartographic models in, 246
  use of multivariate functions in, 260
  use of polygon metrics in, 156
urban geography, example of, 59
U.S. Census Bureau, requesting data from, 112
U.S. Geological Survey (USGS)
  EarthExplorer, 333
  map scales, 28

map symbols, 27
metadata guidelines and examples, 116
use of DEMs by, 82, 83
U.S. Public Land Survey System (PLSS), 35
user communities, as driving force for GIS, 9
user interaction, incorporation of, 286–287
utility maps, 25
UTM (Universal Transverse Mercator), 34, 35
U-turns, 225

V

validation rules, in geodatabases, 92
values, predicting of with interpolation, 191–198
vector data
  basic shapes of, 134
  benefits of raster data versus, 66–68
  building relationships with, 135
  calculating distance along networks, 174
  choosing right source of, 136–139
  as compared to raster data, 133
  defined, 20, 63
  finding features in, 133–150
  getting explicit with, 134
  locating specific features with SQL, 139–144
  optimizing of for speedy searches, 135–136
  predicting outcome, 138–139
  raster data as compared to, 93–94
  searching of with geography, 144–147
  targeting the right data, 136–139
`Vector General` (QGIS), 235
vector model, representing surfaces in, 83–84
vector representation
  adding spatial smarts to GIS with topology, 77–78
  building smarter systems with object-oriented models, 81
  moving beyond simple model, 76–77
  using shapefiles for easy data exchange, 79–80

vector-based model, representation of data in, 65–66
viewsheds, 200–202
virtual reality (VR), 3D GIS and, 338–339
visibility, modeling of with viewsheds, 200–202
visualization, role of GIS in, 10
visualization libraries, 149
visualization methods, 130–131
volume features, 41
volumes
  applying symbology to, 42
  a.k.a. surfaces, 71
  nearest neighbor analysis as effective tool for, 52
  in raster data, 65
  as type of space found on maps, 38, 39
  in vector data, 66
volunteering, 314
voter turnout, analysis of using polygon metrics, 158
VR (virtual reality), 3D GIS and, 338–339

W

Warp (Reproject) tool (QGIS), 258
water flow, characterizing, 205–208
water resource management, ideal data type for, 138
watersheds, 203–204, 208–209
Waze, 103, 109, 226
web applications, GIS-based ones, 285–286
web mapping, 283–287
Web Mercator projection, 33–34, 35
web services
  API as compared to, 296
  getting data through, 297–298
WhiteboxTools, 259
wildlife habitat mapping, ideal data type for, 138
`within`, as spatial operator, 145, 146
Women in GIS, 313

World Geodetic System of 1984 (WGS84), 30, 35, 244
WorldClim, 334

X

X coordinate, as ingredient of surface, 184

Y

Y coordinate, as ingredient of surface, 184, 185

Z
Z value
  displaying and analyzing, 190–191
  examples of, 190
  as ingredient of surface, 184, 185, 188
zero-dimensional, 40
zonal functions, 254–257
zonal statistics, 130
zones, defined, 254
zoom, in interactive maps, 284

About the Author

Jami Dennis, GISP, has been working with GIS for 30 years, blending the art of data visualization with the science of mapping. She's passionate about making data both useful and beautiful, whether it's for designing intuitive maps, crafting compelling visualizations, or uncovering insights hidden in complex datasets.

After 25 years in state and regional government in Arizona, Jami now runs her own GIS consulting practice, helping organizations turn raw numbers into clear, impactful maps and visualizations. She's worked in socioeconomic research, transportation GIS, database development, graphic design, and web development.

Jami is deeply involved in the GIS community, serving on the boards of the GIS Certification Institute, Women in GIS, and the Geospatial Professional Network (formerly URISA). She's also an active volunteer in Arizona's GIS community.

When she's not wrangling data or mentoring new GIS professionals, you can find her camping, hiking, or occasionally staying upright on a mountain bike.

GIS For Dummies, 2nd Edition is her first book.

Dedication

For Dad and Mom, my Hubby, my kids, my siblings, and Tony, who introduced me to the joy of GIS.

Acknowledgments

First and foremost, thank you to Michael DeMers, the original author of *GIS For Dummies.* This second edition builds on the foundation he created. Although I've rewritten the text to reflect current GIS practices, his structure and many of his insights remain at the heart of this book.

I am forever grateful to Hanna Sytsma for giving me the opportunity to write this book. Also, a book like this cannot be written without the support and guidance of a talented team of editors and coaches, so special thanks to Nicole Sholly for introducing me to the *Dummies* style and coaching me through it; to Susan Christophersen for keeping me on track, answering my seemingly endless questions, and boosting my confidence along the way; and to Eva Reid for ensuring that I got my facts straight, lifting me out of imposter syndrome, and reminding me that *data* is plural.

I am incredibly thankful to the many amazing people (far too many to list!) who have supported, inspired, and mentored me, many without even realizing it. You've shown me what's possible when I embrace the superhero within.

Last (but certainly not least), huge thanks to Joni for keeping me sane and to my husband, Michael Lamar, for sticking by my side on long days, late nights, and evening walks.

Publisher's Acknowledgments

Executive Editor: Steve Hayes
Senior Editorial Assistant: Hanna Sytsma
Project Manager: Susan Christophersen
Development and Copy Editor:
  Susan Christophersen

Technical Editor: Eva Reid
Production Editor: Magesh Elangovan
Cover Image: © Ungrim/Shutterstock